VOID

Library of
Davidson College

Bureaucracy and Innovation

Bureaucracy and Innovation

Victor A. Thompson

UNIVERSITY OF ALABAMA PRESS
University, Alabama

Copyright © 1969 by
UNIVERSITY OF ALABAMA PRESS
Standard Book Number: 8173-4812-3
Library of Congress Catalog Card Number: 68-55050
Manufactured in the United States of America

For the older generation, to which the younger generation is passing along its problems.

Contents

Foreword   1

Chapter

    I   Introduction: Organizational Innovation and Organization Theory   3

   II   The Bureaucratic Social System and Innovation   9

  III   Innovation and Organizational Decision-Making   29

  IV   A Program of Research on Innovative Organization   61

    V   Society and Organizational Innovation   89

Appendix   107

Notes   111

Bibliography   139

Index   161

Foreword

IT has by now become almost trite to call attention to the "knowledge explosion." However, while we are frequently reminded that this phenomenon exists, we are seldom told precisely why we should be especially concerned about it, and only rarely are we asked to consider its implications for specific institutions, such as those involved in the administration of large formal organizations. And yet it should be obvious that the "explosion" confronts us with an important question: Are the administrative institutions inherited from a period of information scarcity adequate to the needs of a period of information affluence?

In a period of information scarcity, needs are pressing and not easily satisfied. We might say that man has an excess of ends over means. He must concentrate on overcoming a reluctant nature in an attempt to meet his apparently insatiable needs. He perceives life as a struggle and organizations as weapons in that struggle—weapons made necessary by man's biological limitations.

Under these grim conditions, the functional requisites of "production efficiency"—the key value—determine the nature of organizations and their administration. "Good" administration largely consists of making the organization a more reliable and manipulable "weapon," and the problem of "control" of the organization tends to dominate management's thinking.

The organizational and administrative institutions that have emerged from the period of information scarcity have apparently been fairly well adapted to the demands placed upon them. With increasing efficiency, they have produced a relatively stable product, for a relatively stable market, using relatively stable materials and a relatively stable technology to do so.

Today, however, we have an excess of means (i.e., knowledge) over ends. We have far more information than we know what to do with. Thus, we are seriously in need of creative thinking with regard to values and goals; we need to find new and worthwhile uses for our knowledge.

How well adapted are modern administrative institutions to these innovative needs? That is the basic question to which this book is addressed.

Thanks are due the University of Chicago Graduate School of Business, for providing the opportunity to participate in a weeklong, interdisciplinary seminar on organizational innovation; the National Aeronautics and Space Administration, for financing a research assistant and a reduced teaching load for one semester so that I could explore this subject more systematically; and Professors Robert B. Highsaw and Coleman B. Ransone, Jr., for inviting me to give the Southern Regional Training Program lectures (at the University of Alabama in November, 1967) on which this book is based in part.

Urbana, Illinois   Victor A. Thompson
October, 1968

I

Introduction:

Organizational Innovation
and Organization Theory

THE term "bureaucracy" refers to modern administrative organizations—organizations producing goods or services. In *Modern Organization*,[1] I showed how the qualities of bureaucracy have a two-fold origin. On the one hand, they reflect the growth of technical knowledge and the specialized expertise associated therewith; on the other hand, they result from a particular set of culturally determined and transmitted relations between superior and subordinate roles. A bureaucratic organization is a particular kind of ordered human social system. The study of bureaucratic and other similar kinds of social ordering is called organization theory. In my opinion, organization theory should attempt to explain the part of human behavior that is determined by the social structures that create order within these social systems.

In addition to structure, human organizations have ideologies—bodies of values and beliefs that guide decisions. Ideology distinguishes the human organizations from the beehive and the ant colony. The existence of ideologies leads to the study of decision-making. To study a bureaucratic organization from the standpoint of how decisions are reached usually leads to an emphasis on individualistic psychology or to an examination and critique of the normative decisional rules by which the organization justifies its actions. It is important to keep in mind that this set of decisional rules influences but does not determine decisions. However, decisions are reconstructed *post hoc* in terms of the set of rules. Attempts to study actual decisions empirically have been few and have reflected organization structure only faintly and inefficiently. By and large, they have demonstrated that what decision-makers do and what they say they do are not the same thing. Making such demonstrations with regard to business decisions has been a kind of popular sport for some people in the recent past.

Although decisional theory and organization are two different fields, they are not unrelated. The set of decisional rules accepted by an organization's management will influence policies and procedures and the way the roles within the organization are played. Consequently, I shall be concerned with both organization theory and decision-making theory.

The subject of this book is the relation between bureaucratic structures and innovativeness. There are a number of reasons for being concerned with organizational innovation at this time. In the past, innovation in society took place largely through the birth of new (innovative) organizations and the death of old (traditional) ones. Given the capital requirements of today's technology, this method seems a bit wasteful. We must hope that existing organizations can learn to innovate.

The efficiency of modern credit management would have eliminated the old-fashioned entrepreneur even if the capital requirements of modern technology had not done so. One failure and he is through. (Actually, modern credit management is not quite this efficient, but the phenomenal growth of data-processing technology will soon overcome any remaining defects.) Consequently, we must look for the performance of entrepreneurial functions within the bureaucracy. Whereas failure for yesterday's entrepreneur simply meant the loss of money (someone else's), failure for the modern bureaucrat means the loss of part of his identity. A report of his failure goes into his file—his paper identity, a paper alter ego that follows him inescapably through life—and alters his identity unfavorably. Innovation is more risky for the bureaucrat than for the entrepreneur. Loss of identity is far more serious than loss of money, even one's own. But this observation anticipates our analysis, even as it underscores the importance of studying bureaucracy and innovation.

Other reasons for this perspective can easily be found, even if simple curiosity is not accepted as a sufficient justification. Innovative breakthroughs are obviously needed in such areas as public education and urban government. The swift pace of technological change increases tensions within bureaucracies as the needs of the situation and traditional bureaucratic responses diverge more widely.

Regardless of the reasons for focusing attention on the relation between organization and innovation, the choice of this orientation does not necessarily imply a preference for a high level of innovation over a more stable but productively efficient organization. Both stability and change are desirable under appropriate circumstances. In some matters we want complete predictability and increasing efficiency rather than a high level of change. In other, tradition-ridden areas of life we cry for change, which does not come. The following discussion of bureaucracy and innovation should emphatically not be interpreted as a blanket criticism of modern organizations. However, I would be less than candid if I did not say explicitly that in my opinion most modern organizations in government and business are a bit underinnovative.

At this point I should perhaps give some attention to the problem of defining organizational innovation. By innovation I mean the generation, acceptance, and implementation of new ideas, processes, and products or services. Innovation, therefore, implies the capacity to change and adapt. We can have various degrees of innovativeness, ranging from a capacity to adopt the good ideas of others to the ability to generate and adopt one's own novel ideas. Adoption of ideas new to the organization is quite crucial to our definition. Adoption is necessary to generation. New ideas will not continue to come forth unless there is perceived to be a good chance of adoption. The correlation between suggestions and adoptions in federal suggestion systems is over .8.[2]

This definition is perhaps not specific and operational enough for purposes of quantification, but it is good enough for the heuristic purposes of this book. As a matter of fact, I do not believe that there is yet an operationally satisfactory measure of organizational innovation. I am interested in organizational capacity to adapt, to change, to encourage novel approaches, to release staff imagination, to effervesce with ideas.

Embarking upon the novel involves risk. In a later chapter, where the relation between decisional rules and practices and innovation is analyzed, I shall make much of the element of uncertainty in innovation. Crucial aspects of the innovative process are unpredictable.

Organizations have many outputs besides the production of goods and services. A partial list of outputs would include production values, employee satisfaction, customer satisfaction, profit, community service, survival, and innovation. It has been quite generally assumed in management circles that all of these outputs vary together and in the same direction. This assumption has led to the notion of the "one best way" to organize.

We are beginning to appreciate the fact that these outputs may vary differentially, even inversely. Research on the relation between morale and "output" (a characteristically biased use of the term) lends little support to the notion that these two factors vary together. It has been known for a generation in educational circles that satisfaction with school will not predict school achievement. Conscious decisions with regard to organization structure should be made in terms of trade-offs: so much output gained at a cost of so much employee satisfaction, so much durability at a cost of so much profit, etc. We have very little knowledge of these trade-off functions. There have been few attempts to measure an organization, beyond relating money input to money output. In this book I will attempt to show by means of both analysis and the marshaling of available data that production and innovation vary inversely—the more of one, the less of the other.

In the light of what I have said above, there should be two different but related approaches to the study of innovation—an approach from the standpoint of organization theory, and one from the standpoint of decision-making theory. An organization is a particular kind of (social) system of order. I shall explore whether this particular kind of ordered system provides conditions conducive to innovation. An organization also has an ideology, a cognitive system that rationalizes (read "legitimizes") its past actions and influences its present and future ones. This cognitive apparatus can be reduced to a set of decisional rules. The set of decisional rules that dominates modern bureaucratic action in the West is based on the classical theory of rational choice. Its economic variant is economic rationality.

Classical rationality has been analyzed and criticized by many people and from many perspectives. Without claiming any startling originality, I intend to describe the implications of this set of decisional rules for organizational structure and for innovation. As we shall see, the two are related because the innovativeness of organization is chiefly a function of organizational structure (although personality perhaps contributes some as yet undetermined, but certainly small, proportion of the variance between organizations in this respect).

Classical rationality has been criticized for overlooking limitations of the real world and for being strangely silent on crucial problems of choice (e.g., when to stop searching for consequences). It must be remembered that classical rationality is a normative conception, not an empirical one. The important point is not that people cannot live up to its prescriptions, but that most people think they should and therefore try to do so. Their attempts lead to policies, procedures, and organizational relationships that are incompatible, as I shall attempt to show, with a high level of innovation.

Following the development and presentation of a theory concerning organizational innovation, I will present a concrete program of organizational research based on this theory.

Finally, I will undertake to speculate about changes taking place in our society that will affect the need for innovation and the ability of our organizations to respond to this need. Essentially this speculation will be an evaluation of changes making for and against further bureaucratization of our society.

II

The Bureaucratic Social System and Innovation

IN this chapter I wish to explore the relation between bureaucratic organization as a social system and innovation as one possible output of it. First we must establish the conditions that are conducive to innovative bahavior. Then we will be able to compare these conditions with those provided by the bureaucratic organization.

Although we usually reserve the term "creativity" for the origination of very valuable inventions (novelties of concept) and discoveries (novelties of fact), any adaptive change by an individual or group has the elements of creativity in it—it is new, valuable, and risky. Consequently, our examination must begin with the creative process.

We will first look at the process of individual creativity and then consider the relation of groups to this process. Many empirical studies of the creative process seem to lend support in some measure to Graham Wallas' early formulation of the stages of creativity: *preparation* (assembling the inputs, including the problem), *incubation* (an unconscious or preconscious combining and recombining of internalized components), *illumination* (sudden insight into the solution: "eureka"), and, finally, *verification* (testing and communication).[1] There is still much disagreement about these precise stages, but there does seem to be considerable agreement about two aspects of the process that are particularly important to innovation within bureaucratic organizations. First, the creative process is an irregular one, and it often seems aimless and unpredictable. It is characterized by sudden leaps. From the point of view of production norms, it seems inefficient. It does not seem to be a disciplined, diligent, pay-attention-to-business sort of thing. Second, the creative process is characterized by slowness of commitment, by suspended judgment, by refusal to grasp the opportunity, by refusal to make quick, decisive judgments. It is inclined to make a painfully full exploration at the initial analytical stage and to continue search long after satisfactory solutions have been found.[2] As we shall see later, these qualities of the creative process pose a conflict between an interest in innovation and an interest in production.

Empirical studies of creativity suggest certain conditions conducive to creativity that are of especial relevance to bureaucratic organizations.[3] These conditions are: (1) psychological security and freedom, (2) a great diversity of inputs, (3) an internal or personal commitment to the search for a solution, (4) a certain amount of structure or limits to the search situation, and (5) a moderate amount of benign competition. The individual needs to be personally secure even though the problem poses a subject-matter uncertainty.[4] Failure should not be a personal catastrophe nor should sheer success in solving the problem be of overriding importance, since either or both of these conditions might cause the individual to settle for the first satisfactory solution he comes upon and keep him from continuing to search for an even better or more novel solution.

To be creative, the individual needs a richness of experience with the subject matter; he needs a great diversity of inputs, of ideas, of stimulation. It is often observed that serendipity comes to the individual who is prepared to use it. It is not all luck. As for motivation, it appears that high internal commitment to the search, short of debilitating anxiety, is the ideal condition; rewards are intrinsic rather than extrinsic, so that to a considerable extent the process of solving the problem is an end in itself. To the extent that rewards are external, it appears that they take the form of improved esteem in the eyes of similarly committed peers rather than an increase in interpersonal power relative to peers or a mere improvement in income as such. Thus, a certain amount of benign peer-group competition seems to be conducive to creativity. Finally, we should point out that our emphasis on freedom and security does not signify an absolutely unattached and free-wheeling individual. The myth of the inventor as a lone genius does not get much support from the empirical data on creativity. A certain amount of external direction and limitation of activity in the form of concern and evaluation is actually conducive to creativity. As one writer has put it, if the creative thinker's environment is not somewhat limited by others, one of the first things he must do is limit it for himself.[5]

In sum, the empirical conditions for individual creativity suggests a sort of golden mean: some freedom, but not too much; high internal commitment to the task, but not too high a commitment; a high proportion of intrinsic rewards, but some extrinsic rewards as well; some competition, but not cutthroat, winner-take-all competition.

We now turn our attention to the effect of group experiences upon creativity. The literature on problem-solving suggests that for some kinds of tasks groups are more effective than individuals, while for other tasks the reverse is true.[6] However, the individual is always embedded in a pattern of group relations, and this entanglement affects all of his activities, creative or otherwise. Here we note certain effects of the group on the individual that have some bearing upon our problem of evaluating the innovativeness of bureaucracy.

When we speak of groups we are referring to spontaneous role relations arising out of a great deal of face-to-face interaction, rather than more formal and structured role relations such as those between superior and subordinate. We are interested in the effects of the group both when its members are physically together and when they are not.

The group can provide protection for the individual and thus contribute to his security and freedom in a number of ways. It can give him support so that he does not stand alone with his novel ideas before the world. It can both correct his errors and indicate approval when he is on the right track.[7] It can share responsibility and thus reduce the riskiness of promoting new ideas. A group can also help round up the support needed to get new ideas accepted and implemented.

The group can also provide great diversity and richness of inputs, so important in stimulating creativity. In this connection, it is probably important that the group represent various specialties and not be composed of a number of identical subject-matter specialists. New ideas of sufficient complexity to go beyond the area of one person's expertise must be, in some sense, group products. Furthermore, in the group with varied skills, someone will at all times find himself in the position of being nonexpert, a layman in the subject under discussion. His very ignorance and naïveté allow him to ask questions that, because they are from a fresh perspective, might never occur to the experts. We would like to call this group role the "generalist role." It seems to have been given very little attention in studies of group problem-solving.

Whereas the "generalist" as a universal omniscient genius is a figment of the imagination (in recent centuries, anyhow), and the "generalist" as a superficial dilettante with a little knowledge in many fields is unimportant, the generalist *role* in problem-solving may be a very important one. It may help us to reduce the conservative, conformist effects of what Thorstein Veblen called "trained incapacity." It deserves further study. While no one *is* a generalist, anyone can play a generalist *role*.

The group can provide some of the extrinsic rewards of growing esteem, and it can provide the arena for the benign competition mentioned above. To serve in this latter capacity, the group needs to be large enough for an individual in any specialty to find peers in that same specialty. If one individual is the sole expert in his field, he will not have competition. (Also, the supportive and error-correction functions of the group will not operate with respect to him.) For the same reason, the group must be free of rigid stratification. Competition between strata is not permissible. (Stratification actually creates several groups or subgroups with little communication between them. Consequently, a rigidly stratified group, from our point of view, is a contradiction of terms.)

It has been suggested, in this connection, that the most brilliant member of the group should not be the superior officer. Putting the most brilliant person in the superior's position reinforces tendencies toward stratification. If this suggestion is sound, it creates an interesting problem. Our tendency to define success in terms of moving up the managerial hierarchy and our commitment to achievement norms mean that the "best" person must always be placed in the position of superior (or that the superior must be judged the "best" person in the group). If we follow the suggestion above, we do *not* place the "best" man in the superior's position, and this forces us to give up either our definition of success or the norm of achievement. (Here we have already reached the suggestion that innovative organization is pluralistic; one aspect of pluralism is to have more than one criterion by which such terms as "best" are defined. More of this later.)

It seems to be quite clear that competition between groups is a very real force in individual behavior. Individuals often become identified with the groups in which they work so that group goals become personal goals. Thus, two or more groups of interdependent people working on the same problem, but possibly going in different directions or by different paths, can stimulate each other to find better or more novel solutions. Such a situation also improves the chances of finding outstanding or novel solutions in less time and with less expense, as has been demonstrated in the field of new weapons development.[8]

Before leaving the subject of the group's effect on creativity, it is necessary to mention a possible negative effect. Many studies attest to the fact that groups, over a period of time, exert powerful conformist pressures on their members. The group generates norms and standards of many kinds and powerful sanctions to assure compliance with them. Consequently, there is nothing about groups as such that can be guaranteed to increase individual creativity; they could, in fact, have just the opposite effect. However, if a group has innovative norms, the powerful conformist pressures will strongly reinforce the other group aids to innovation discussed above.[9] We should also mention the possibility that a group's norms may be irrelevant to individual creativity and have no effect on it in any way. Recent evaluations of "brainstorming," for example, conclude that this method of group problem-solving is less efficient than individual problem-solving in generating new ideas.[10]

In this chapter we want to assess the effect of organizational structures on innovativeness. These structures are only to a modest extent planned and rationalized relationships. Mostly they are heritages of the past, acquired in the same way that family organization is acquired.

Before discussing a particular kind of social system—namely, bureaucracy—it would be well to say a few words about social systems in general.

A social system determines its members' (participants') behavior to the extent that their behavior is related to the system. In many ways these system control processes must be latent in order to work. Knowledge of them would change the processes, hence the social system, either because members would exploit the processes or because they would resist them. At the very least, knowledge of them would produce system instability pregnant with unpredictable change. Thus, the members of a social system must remain largely ignorant of its latent functions and processes. Otherwise, the system changes and reestablishes control in new ways, and new latent functions and processes emerge. Only a social system that could work without participants could avoid this. There is an element of indispensable mystery about going concerns.[11] Consequently, innovation is limited by the functional processes of the surviving system. There is a limit to the amount of change a system can absorb without dissolution. We are interested in the specific restraints on change to be found in bureaucratic organizations.

Modern bureaucratic organizations are framed around a powerful organization stereotype. Following Max Weber, we propose to call this stereotype the "monocratic" concept of organization. It seems reasonable to assume that this stereotype reflects conditions prevalent in the past. Important among these conditions, because they no longer hold, were: (1) great inequality among organization members in social standing and abilities and a corresponding inequality in contributions and rewards, and (2) a very simple technology, readily within the grasp of single individuals. Out of these conditions, among others, there arose the monocratic stereotype and the roles through which it is realized.

According to this stereotype, the organization is a great hierarchy of superior–subordinate relations in which the person at the top, assumed to be omniscient, issues the general order that initiates all activity. His immediate subordinates make the order more specific for their subordinates, the latter do the same for theirs, etc., until specific individuals are carrying out specific instructions. All authority and initiation are cascaded down in this way by successive delegations. There is complete discipline, enforced from the top down to make certain that these commands are faithfully obeyed. Responsibility is owed from the bottom up. Reports on the carrying out of orders and the results obtained flow upwards to the top where they are compared with top management's intentions. As a result of this comparison, orders are modified and again flow down the line, and the cycle is repeated. The organization is perceived to be a feedback loop.

To assure predictability and accountability still further, each position is narrowly defined as to duties and jurisdiction, without overlapping or duplication. Matters that fall outside the narrow limits of the job are referred upward until they come to a point where there is sufficient authority to settle the matter. Each person is to receive orders from, and be responsible to, only one other person—his "superior."[12]

Clearly the organization is conceived to be the tool (or weapon) of an "owner," and the "owner" is duplicated on a smaller scale at each hierarchical level. Such a system is monocratic because in any conceivable situation there is only *one* point or source of legitimacy. Conflict, though it may occur because of the weakness and immorality of human beings, cannot be legitimate, and so the organization does not need formal, legitimate bargaining and negotiating devices. Thus, while it may be empirically more fruitful to conceive of the organization as a coalition,[13] according to the monocratic stereotype the organization, as a moral or normative entity, is the tool of an "owner," not a coalition. Coalition and other conflict-settling activities, therefore, take place in a penumbra of illegitimacy.

The inability to legitimize conflict depresses creativity. Conflict generates problems and uncertainties and diffuses ideas. Conflict implies pluralism—a dispersion of legitimate power. Thus, it necessitates coping and searching for solutions, whereas overriding, concentrated power or authority can simply ignore obstacles and objections. Conflict, therefore, encourages innovation. Other things being equal, the less bureaucratized (monocratic) the organization, the more conflict and uncertainty and the more innovation.[14]

The modern bureaucratic organization is dominated by the monocratic stereotype. The stereotype is institutionalized in a system of boss-man roles. Modifications of the stereotype, therefore, come through variations in individual role-playing. The monocratic organization stresses hierarchical authority and communication ("going through channels"). It is highly oriented toward control, predictability, reliability. Control is facilitated by defining jobs narrowly, programming activities into routines to the greatest extent possible, and fixing responsibilities by avoiding all overlapping and duplication. The monocratic stereotype dictates centralized control over all resources. It stresses iron discipline from the top down, enforced by the centralized administration of extrinsic rewards and deprivations. It can control only through extrinsic rewards such as money, power, and status, because it demands the undifferentiated time and effort of its members in the interest of the "owner's" goals. Even as the organization is a tool, so are all of its participants. There is no place for "joy in work." To admit the propriety of joy in work would be to admit an interest other than the "owner's" and to lose some control over the participants. Carried to its logical extreme, the only person in the monocratic organization who could innovate would be the "owner" (the "top man").

The necessity of relying upon extrinsic rewards of money, status, and power forces the organization to barter off its hierarchical or managerial positions as rewards for docility and compliance. The organization has to "consume" itself, by giving parts of itself away as rewards. Such a reward system depends upon the organization's ability to find substantial numbers of people who think that the exchange of their undifferentiated time and effort for a poor chance at a small and rapidly diminishing group of positions is a good bargain. It is doubtful that this would have been possible without help from other social institutions, including religious ones. The general belief that work is not supposed to be enjoyable has helped. So has the social definition of "success" as moving up a managerial hierarchy. The further belief that the good man is the successful one has closed the system.

As education has become accepted as a criterion of social class, the blue-collar group and a large part of the lower white-collar group have been eliminated from the competition for these great, scarce status prizes. Furthermore, it seems to be true that more and more highly educated people are rejecting this bargain and seeking basic need satisfaction outside the organization—in hobbies, community activities, and their families.[15] Consequently, organizations are increasingly hard pressed to find rewards sufficient to induce the needed docility. Whereas the use of money alone has raised the price of goods and perhaps priced us out of some markets, it does not seem to have been an outstanding success in promoting output.[16] Furthermore, to protect the soundness of its most powerful currency—hierarchical positions—the modern organization has had to undervalue nonhierarchical activities.

Today, with the enormous expansion of knowledge flooding the organization with specialists of all kinds and making the organization increasingly dependent upon large numbers of people with years of pre-entry preparation, this reward system is facing a crisis. The person with a great amount of relevant pre-entry training finds that he can "succeed" only by giving up work for which he is trained and entering work—management—for which he has had no training.[17] And since in an open society "success" must be available to all, management (hierarchical) positions cannot be restricted to a specially trained group of people.

The extrinsic reward system, administered by the hierarchy of authority, stimulates conformity rather than innovation. As remarked above, creativity is promoted, for the most part, by an internal commitment, by intrinsic rewards. The extrinsic rewards of the esteem of one's colleagues, and the benign competition through which it is distributed, are largely foreign to the monocratic, production-oriented organization. Hierarchical competition is highly individualistic and malignant. It does not contribute to cooperation and group problem-solving. As we shall see below, it establishes an organizational political system that is extremely resistant to innovation.

For people committed to this concept of success, of self-evaluation, with its malignant competition, the normal psychological state is one of anxiety, in some degree. This kind of success is dispensed by hierarchical superiors. Furthermore, the more success one attains, the "higher" one goes, the more vague and subjective become the standards by which one is judged. Eventually, the only safe posture is complete conformity.[18] Innovation is hardly even thinkable under these conditions. To gain the independence, freedom, and security required for creativity, the normal individual has to reject this method of determining his worth, this concept of success. But even those who have adopted a different life pattern and measure their personal worth in terms of professional growth and the esteem of professional peers must feel a great deal of insecurity within these monocratic structures, because the *opportunity* for growth is under the control of arbitrary authority, and so, especially, is the work they are asked to perform. The organization claims a right to their undifferentiated time and effort, and though it is not likely to press this claim to the extreme (for example, by asking engineers to do janitorial work), the production ideology of the monocratic, tool-organization characteristically cuts the job somewhat smaller than the man.

The hierarchy of authority is a procedure whereby organizationally directed proposals from within are affirmed or vetoed. That is to say, it is the procedure whereby legitimacy is dispensed, whereby ideas become official. It is a procedure that works in such a way as to give the advantage to the veto. The reason for this result is that monocratic systems do not provide for appeals. An appeal implies conflicting rights that must be adjudicated. In fact, however, the superior's veto of a subordinate's proposal legitimately ends the matter. On the other hand, if the organization's resources are involved, an approval must usually go higher, where it is again subject to a veto, because of the common pattern of centralizing control over resources. Thus, even if the monocratic organization allows new ideas to be generated, it will probably veto them. (Actually, they go into files and wait a new chance, which may be generated by a crisis, or a catalyst such as a consultant.)

As we shall see in the next chapter, the modern bureaucratic organization is strongly oriented toward production interests. Production interests lead to overspecification of human resources. In the past, this has resulted in organizations composed of unskilled or semiskilled employees, both blue-collar and white-collar. The jobs of these people were to carry out more or less simple procedures devised within the organization that hired them. No special previous preparation (beyond reading and writing) was required, and any reasonably intelligent person could become proficient in his job in a short time.

The gradual exclusion of blue-collar workers from the success system has alienated them, to the effect that these organizations are divided into "management" and "labor." Blue-collar workers have sought protection in unionism. The white-collar unskilled have adapted individually to their fate. Being within the success system, each office worker felt he had a chance to "get there." Their adaptations did much to set the tone and style of the bureaucratic organization. For this reason, the following analysis will concern only the white-collar unskilled (or semiskilled), the office worker, the desk worker.[19]

The important point about the desk classes is that their work is determined by the organization rather than by extensive pre-entry preparation. The interorganizational mobility of the desk classes declines with time. They become dependent upon the organization for status and function—in the extreme case, for everything that is worthwhile. If they do not become alienated, they become organization men, loyal to the organization that supports them, thereby strengthening the system of organizational authority. Deprived of intrinsic rewards related to the work or the rewards of the growing esteem of professional peers, they become largely dependent upon the extrinsic rewards distributed by the hierarchy of authority, thereby greatly reinforcing that institution. Their dependence upon organizational programs and procedures for whatever function they acquire induces a conservative attitude with regard to these programs and procedures. They may even hypostatize them into "natural laws," losing sight of their purely instrumental significance.[20]

Except for that small minority who appear likely to win one of the few great status prizes, the state of morale of the desk classes is one of chronic, though not necessarily extreme, dissatisfaction.[21] Overspecification plus dependence upon extrinsic rewards of promotion result in vast overrequirements of qualifications. In most cases, the individual becomes qualified for the minor incremental increases in difficulty of the next higher job years before it becomes available. In the meantime, he has only his salary, his hopes, and perhaps, if he has pleasant working companions, a few social rewards. The pretense that his promotion has come about because he is "now ready" for the "more difficult" job only makes his plight the more poignant.

The resulting, easily recognized mental and emotional condition has been called the "bureaucratic orientation."[22] The monocratic, production-minded organization tends toward this orientation.

In the absence of a management profession, for reasons outlined above, management must be regarded as a desk-class phenomenon. It is significant that in business organizations almost half of the higher management comes from the desk classes and only a fourth from the renegade professionals. (In the federal government, these proportions are almost exactly reversed.)[23]

The bureaucratic orientation is conservative. Novel solutions, using resources in a new way, are likely to appear threatening. The bureaucratic orientation is politically minded. It is more concerned with the internal distribution of power and status than with the accomplishment of the organization's goals. It converts the organization into a political system concerned with the distribution of extrinsic rewards.[24] The first reaction to new ideas and suggested changes is most likely to be: "How does it affect us?" Some observations of the decision-making process in business organizations suggest that the search for alternatives and consequences in these organizations is largely an attempt by political groupings to find answers to that question—"How does it affect us?" The expectations of consequences upon which decisions are based are heavily biased by political interests.[25] Since there is no consensual or objective basis for distributing the extrinsic rewards of the bureaucratic organization, this political process is a functional necessity.

There are many lines of defense, and if new activities cannot be blocked entirely, they can at least be segregated and eventually blocked from the communication system if necessary. Typically, the introduction of technical innovative activities into modern organizations is accomplished by means of segregated units, often called R and D (Research and Development) units. Segregating such activities prevents them from affecting the status quo to any great extent. The organization does not have to change. It can merely add another unit and thus "cover" the subject. The R and D unit is often an isolation ward where creative people are kept until they become domesticated and can be put to work servicing the needs of the units already on the scene—that is, servicing the status quo.[26]

We should add that it is not only the organizational political system that causes the segregation of new problems. There is often no place in the existing structure into which they can be fitted. When a new problem appears, the monocratic production-oriented organization is likely to find that the resources of authority, skills, and material needed to cope with it are unavailable because they have already been fully specified and committed to other organization units. Since no existing organization (jurisdiction, unit) has either the authority or the uncommitted resources to deal with the new problem, a new organization unit must be established.

If an organization takes its innovation needs seriously and tries to meet them through a segregated R and D unit, it is likely to run into a great deal of trouble trying to stimulate innovativeness within the segregated unit. If, as we have argued, it cannot use the extrinsic reward system upon which the political system is based, it must fumble toward a reward system that is entirely unnatural to the monocratic organization. It must establish conditions that are entirely foreign to the conditions of production upon which the monocratic organization is based. Two milieus, two sets of conditions, two systems of rewards, two organizations, must be established, one for innovation, the other for the rest of the organization's activities.

This duality is dangerous. Not only is it divisive and upsetting to the existing distribution of satisfactions, but it actually undermines the legitimacy of the monocratic system itself. If the organization's success is seen to be dependent upon activities that can only be well performed within a nonmonocratic social system, why is the same not true for all other activities of the organization? A truly innovative R and D unit introduces a corrosive pluralism into the monocratic structure. For this reason, one suspects that R and D units are rarely allowed to be truly innovative, and it has been reported that unhappy R and D personnel are sometimes bought off by top management with high salaries and fancy titles.

Beyond the political interests in the distribution of extrinsic rewards, there are other factors that strengthen tendencies toward parochialism. The organization seems to factor its activities into narrow, single-purpose, exclusive categories and to assign these to subunits composed of a superior and subordinates. An effort is made to see that all activities are assigned so as to avoid any overlapping and duplication. Both the nature of the resulting work and the obligations of subordinates to their superior tend to restrict communication to that with peers and a common superior.

Very often strong personal identification with the particular subunit and subgoal develops within this pattern, so that members of any one unit know little about what other units are doing—and care less. The needs of other units are not attended to and may not even be perceived.[27] The organization tends to become a collection of small entities with specific boundaries and frontiers. When work is completed in one entity, it is "handed over" to another and interest in it is dropped; from now on, it is "somebody else's baby." Interest tends to be not in the work but in protecting the records and protocols of the hand-over transaction, so that blame, if forthcoming, can always be pinned on another unit. The unit to which the work is handed has an interest in discovering flaws in it so that any future difficulties can be blamed upon the unit from which it received the work in the first place.[28] In this way the tough, impersonal problems of the organization tend to become personalized matters of "the good guys and the bad guys," or "the smart guys and the dumb guys," and consequently the problems are never faced at all.

Although tight, narrow, exclusive mission assignments are justified as needed "to pinpoint responsibility," they actually encourage irresponsibility in regard to new problems and ideas. They facilitate "buck-passing." It is always possible for any individual (or unit) to say: "This is not my problem; it does not fall within my jurisdiction; let somebody else take care of it."[29] It is becoming increasingly evident that a somewhat more untidy structure, one with overlapping and duplicating efforts to solve the same problem, achieves more outstanding and novel solutions, in less time and with less expense, than does the tidy monocratic structure with "responsibility pinned down" in exclusive jurisdictions.[30]

Present methods of departmentalization, of assigning missions to subunits, also encourage parochialism. At the simple unit level (superior and subordinates), it is often, but not always, an aggregative grouping—a number of people with the same skills doing the same thing. "Like activities should be in the same place." Lacking the stimulation of different skills, views, and perspectives, and the rewards of project completion and success, such groupings are likely to pin their hopes on extrinsic rewards and seek them through the organizational political system. Assuming that any enthusiasm and interest are generated at all, they are more likely to be spent in playing politics than in creative activities.

Other simple units, even though not composed of aggregations of people doing the same thing, are very often composed of overspecified desk classes carrying out some continuing program—say getting out the house organ, or recruiting, or keeping stores. In such an integrative grouping there may be more interpersonal stimulation, but overspecification—the sheer subprofessional simplicity of the jobs—prevents the diversity and richness required for anything but very minor innovations. Furthermore, the loyalty and attachment of the desk classes to their hypostatized programs makes them generally conservative and resistant to new ideas.

The aggregative grouping has neither interdependence nor goal. Group innovation is therefore impossible. Individual innovation in the interest of the organization is hardly likely, unless the organization offers rewards for it. Some organizations reward individual innovative suggestions through a suggestion-box system, but such systems are rarely successful. As far as aggregative units are concerned, the lack of diversity among the inputs to these discrete, noninteracting individuals, especially if the units are composed of overspecified desk workers, diminishes the likelihood of any important innovative insights. For integrative units, suggestion boxes are frequently disruptive because the true authorship of the suggestion is likely to be in dispute, and the group will often feel that the idea should have been presented to it rather than individually presented for an award.[31]

One final aspect of monocratic organization should be evaluated from the point of view of its impact on innovation. I refer to the theory of monocratic responsibility. According to this theory, praise and blame attach to jurisdictions. Extrinsic rewards are to be conferred or denied, and even withdrawn, according to the successes and failures that occur within jurisdictions. Since all possible activities are thought to be distributed in exclusive fashion to exclusive jurisdictions, it should always be possible to find some individual person to blame when anything goes wrong. People are to be punished for mistakes as well as wrongdoing, and they are to be punished for failures within their jurisdictions whether or not the failures are caused by their own activities. ("He should have prevented it. It was his responsibility.")[32]

Though this theory is no longer strictly applied, it is still feared. Thus, an individual may hesitate to advise an organization to take a particular action, even though he has very good reason to believe that the likelihood of a satisfactory outcome is high. Should the action fail, he may be a personal failure. It is difficult to apply the concept of probability to personal failure. Probability applies to a series of similar events like the tossings of a coin. However, one feels, rightly or wrongly, that he can only fail once. Therefore, what would be rational from the standpoint of the organization's goal may appear irrational from the standpoint of the individual's personal goals. Consequently, the good advice may not be given to the organization.[33]

New ideas, departures from the tried and true, are particularly speculative and hence particularly dangerous to personal goals, especially to the extrinsic personal goals of power and status. Consequently, the monocratic organization, structured around such extrinsic goals and explicitly committed to this stringent theory of responsibility, is not likely to be highly innovative. And that part of this kind of organization most committed to such goals and to this theory is precisely the managerial hierarchy. Here "responsibilities" weigh heavily. Therefore, the generation of new ideas is not likely to be a hierarchical phenomenon, unless unusually risky persons have found their way into some of the hierarchical roles. The same considerations apply with regard to the acceptance and implementation of new ideas. Old veterans of the organizational game will tell you that it is better to go by the book, avoiding risky innovations.

Empirical evidence that different kinds of structures are optimal for different kinds of problems is quite compelling.[34] Almost equally compelling is the evidence that leadership role assignments need to be changed as the situation changes.[35] Bureaucratic rigidity makes such rational structural alterations all but impossible. We cannot escape the conclusion that our current organization structures are *not* the most rational adaptations for *some* kinds of problem-solving at the very least. While it is true that experimental groups have been successfully restructured from bureaucratic to collegial by means of verbal redefinitions of roles along lines perceived to be more appropriate to the task at hand,[36] such restructuring is probably impossible in real life, tradition-bound organizations as presently constituted.

Before we leave the evaluation of the monocratic organization structure and go on to a consideration of bureaucratic decision-making, we should mention briefly a monocratic variant which is highly innovative—in the short run. New organizations are sometimes begun by highly creative individuals who attract like-minded people, maintain an atmosphere conducive to innovation, build up a powerful *esprit de corps* and high morale, and achieve a very high level of organizational creativity. Often these are small engineering or research organizations founded by a competent engineer or scientist, assisted by a small group of able peers loyal to him personally. Such organizations are new and small and not yet bureaucratized, and many able young people may be attracted to them because of the opportunities they provide for professional growth.

As these organizations grow larger, however, and particularly after the charismatic originator is no longer there, the monocratic stereotype asserts itself and they become bureaucratized. This phenomenon is an ancient one, discussed by Max Weber as the "institutionalization of charisma." One manifestation of it is seen in the post-revolutionary bureaucratization of successful revolutionary organizations. Thus far we have not been very successful in preventing it.

III

Innovation and Organizational Decision-Making

IF an organization, or any group, is to act as an entity, it must have a body of doctrine, an ideology, that explains what it is doing and what it ought to do. For the modern bureaucratic organization, this body of doctrine could be called, in shorthand fashion, a "production ideology." It conceives of the organization as having an "owner" who has a goal, or preference ordering, that he wishes to "maximize" by means of the organization. The organization is a tool (or weapon) for reaching this objective. The various participants are given money in return for the use of their undifferentiated time and effort as means for achieving the "owner's" goal. As Henry Ford said: "All that we ask of the men is that they do the work which is set before them."[1] "Management" consists of functions and processes for perfecting the "tool." Management's interest, therefore, is control of all intraorganizational behaviors so that they become completely reliable and predictable. From the standpoint of this production ideology, innovative behavior can only be interpreted as unreliability.

While actual behavior in organizations reflects this ideology very imperfectly, the ideology does not seem to have been solidly replaced by any other as yet; evaluations of organizations and of parts of organizations are still predominantly in terms of it. Although, as

Cyert and March rightly show, it is more fruitful for empirical studies to regard the organization as a coalition, evaluations are still largely couched in terms of how well the organization "maximizes" the "owner's" goals.[2] For example, it is possible to interpret the evaluation, "inefficient," as a criticism of the distribution of the "slack" within the organization (surplus satisfactions beyond those needed to induce needed contributions). It often signifies the critic's distaste for the fact that all slack has not been collected and paid to the owner. The "efficient" organization would be the one in which no one but the "owner" received satisfactions, in any form, beyond what was needed to induce his contributions. For although the various satisfactions (payments) received by individuals are not measurable upon a common scale, it is usually possible to exchange them for *some* amount of the "owner's" values. Thus, the delight a person gets from his status could be converted into money by removing all status symbols from his office and selling them. The delight an engineer gets from his many skills and large amount of knowledge could be converted into increased production, control, and predictability by restricting his activities to the design of tool handles.

The production ideology leads to rapid and detailed specification and commitment of resources. Of special interest to us is the detailed specification of human resources. Adam Smith's advice, that the job of pin-making should be reduced to that of making a part of the pin, has been generally followed. We will refer to this response as the "Smith's pins" effect. It has been said that the detailed specification of human resources reduces investment costs per unit of program execution (presumably investment in training).[3] The production ideology usually results in jobs that are smaller than the men who fill them, whether the men are industrial workers turning nuts all day long or engineers drafting tool handles.[4] Consequently, we will refer to the detailed specification of resources, somewhat argumentatively, as the "overspecification" of resources.

The relation between production interests and the overspecification of resources involves more than Smith's admonition. The more specific activity is also a more measurable one and hence a more controllable and less risky one. The more broadly the job is defined, the less operational, the less controllable, and the more risky it is. The returns are less certain. The current decline of overspecification does not contradict these statements. It is attributable to the fact that current technology cannot tolerate the former degree of overspecification, and to the fact that the professional interests of more and more collaborators in the organizational coalition demand side payments in the form of opportunities for professional growth. There may also be a growing felt need for innovation, and overspecification denies the diversity of inputs that innovation requires.

Interest in predictability and control lies behind the tendency to program more and more activities within the organization. If a person's activities are even partially unprogrammed, he is partially out of organizational control and under self-control. He may even be engaged in innovation or search leading to innovation. (He may also be loafing.) Many important innovations have resulted from "illegitimate" activities contrary to the express orders of management (e.g., the disc memory unit of the random access computer). To some extent, control-oriented management is defined in terms of worrying about these nonprogrammed spheres of activity. Such management must breathe easier when everyone is at his desk or bench busily "working."[5] Furthermore, the organization is more a manipulable tool of the "owner" when everyone stays within reach of his telephone (or foreman) and is not wandering about "communicating" and "collecting ideas." From the standpoint of a thoroughly control-oriented management, communication should only occur in official meetings or interviews. Given this need for control, the new data-processing technology will bring great pressures for more control and more centralized control. This prospect bodes ill for innovation.[6]

Production orientation, the need for measurable controls, and the overspecification and rapid commitment of resources greatly reduce the innovative capacity of an organization. The overspecification of human resources sacrifices individual innovative capacity by greatly reducing input diversity. The narrowness of the job greatly reduces inputs from the environment, including inputs of problems, and the nonuse of skills and knowledge reduces internal inputs and diversity. The generally trivial output of suggestion-box systems illustrates the effects of these conditions. Furthermore, highly specified and fully committed resources leave neither the time nor the free resources required for innovation. For example, a man on a highly specified and hence measurable job can be given, and will be given, production goals that leave him no time for innovation. Innovative activity is not measurable and predictable; it is not "efficient;" it is likely to look like loafing. Hence, under the production ideology it must be stamped out. In a strict sense, only the "owner" can legitimately innovate.

The sacrifice of innovative potential through Smith's-pins overspecification has been increasing with the advance of education in this country. Even as we overspecify jobs, we "overrequire" as to qualifications for these jobs. We want a high school diploma for a job that could be performed by a chimpanzee. We want college degrees for the desk classes doing work requiring only general intelligence and reading, writing, and sometimes arithmetic. The amount of engineering knowledge required for many engineering jobs could be learned in high school. We like refined waitresses and educated clerks. Overrequirement (relative to overspecification) may be greater in government, but it is rather obviously an enormous drain on social capital in both government and business.[7]

The solution, of course, is not to cut back education but to realize a better return on our educational investment by enlarging jobs rather than throwing away a large part of the educational product as is done today.

Production-oriented management proceeds under a monocratic intellectual apparatus, referred to by economists as economic rationality. Charles Lindblom, a critic, calls it synoptic rationality.[8] It urges comprehensive problem-solving, "getting all the facts," complete analysis and understanding of a problem. It posits a single goal set or preference system (the "goal" or "end" or "objective") and demands that all alternate approaches be canvassed and evaluted in terms of all of the consequences of each, and that the particular alternative be chosen that is best in terms of the single goal set.[9] For organizations, the monocratic single goal set is achieved by assuming that the organization has an owner; the goal set is his, and all other values are irrelevant (actually, they are supposed to be neutralized by the payment of wages, or the denial of individual rights).

This intellectual apparatus has a number of problem-solving rules that require marginal comparisons of all possible activities. They require, therefore, constant reexamination and search for new alternatives and consequences as conditions change, and constant reinvestment of resources according to marginal advantages. Optimization of the goal set requires balanced progress, and this requires complete coordination of all activities. Overlapping and duplication are considered a waste of resources. To eliminate variances due to human defections, there is a strong tendency toward mechanizing or programming all activities. The organizational structure most consistent with this intellectual apparatus is Max Weber's monocratic bureaucracy.[10] Classical rationality encourages strict, hierarchical, authoritarian role-playing within the bureaucratic organization.

This intellectual structure is designed to work with certainties or determinate probabilities, and it becomes increasingly inappropriate as the amount of uncertainty increases with regard to inputs and outputs. For example, the subrule governing search for alternatives and consequences says that one should continue searching until the marginal cost of search begins to exceed the marginal improvement in alternatives found.[11] But this rule obviously assumes that one knows the *value* of what he is looking for, and hence knows *what* he is looking for. The rule is not meant to govern "idle curiosity."[12] The uncertainty of means and ends, of inputs and outputs, increases as we proceed along a continuum from production to development to applied research to basic research. What we wish to show, then, is that the monocratic intellectual apparatus and its Weberian organizational counterpart are increasingly inappropriate as we go from production, at one end of our continuum, to basic research at the other, and that this means that the intellectual and organizational models appropriate for production are different from those appropriate for innovation.

We feel, intuitively, that there must be some relation between input and output in knowledge-producing activity, in R and D, but no one has yet demonstrated a "production function for invention." All that past studies do is to indicate that R and D expenditures "seem to accomplish something."[13] As economists have shown, technical progress has substituted labor-saving capital for increasingly expensive labor. In production, the element of unit cost that has been reduced the most over the years has been the cost of the factor that has increased most in price—namely, labor. But these vague general facts do not establish a production function for R and D expenditures. They are, in a sense, tautological. Technical development, by definition, is the progressive substitution of nonhuman power and activity for human, and the rising cost of labor simply reflects the greater output of machines and hence the greater per capita incomes.

The individual organization has no way of knowing whether

a particular investment in R and D will pay off, how long it will have to wait for the payoff, and whether or not the benefits will be largely appropriated by others. Reports differ, but many, many years elapse between the beginning of R and D projects and their successful exploitation, and perhaps as many as 80 per cent of them are never profitable.[14] It is seldom possible for the organization that takes the risk to appropriate all the benefits, and the benefits increase with time, which makes the later, exploitive stages more attractive than the earlier, inventive ones. Most American companies plan to get in during the later stages, leaving invention or development to someone else.[15] Premium prices and profits are possible for only a short period and so the tendency is for rapid closure and commitment of resources rather than the full initial exploration and slow commitment that is characteristic of creativity.[16] In short, the closer we come to basic research, the more irrelevant are the norms of economic rationality.

The production ideology raises questions about the legitimacy of innovation. It does not seem to be possible to know, or even expect, that the "owner's" values will be promoted by a true innovation (real novelty). It is not possible to know how much in the way of resources should be allocated to search when you do not know the value of what you are looking for—when you do not even know what you are looking for. It would seem that the model of "rational behavior" cannot account for invention, and that invention is a *nonrational* process.[17] This dilemma poses the question of the legitimacy of allowing innovation for its own sake and helps us understand why innovation, considered from a psychological as opposed to a moral or rational point of view, is more likely to occur in an organization with considerable slack—one in which the organizational achievement level has been substantially above the aspiration level. Though innovation for its own sake is no more legitimate in such an organization, there is likely to be less pressure to marshal all of the "owner's" resources in the pursuit of the "owner's" goals as currently understood. In many large industrial research departments today, research workers are allowed to devote some part of their time, often 10 per cent, to research into whatever interests them at the moment.

Studies of organizational innovative decisions show that they do not follow the prescriptions of the comprehensive rational model. Business firms that automate do not "get all the facts." For the most part they simply weigh reduction in manufacturing costs (chiefly labor) against increased investment, though sometimes the likelihood of increased maintenance costs is also recognized.[18] Once the decision to invest in new equipment is made, it is common for the organization to plunge nonrationally, to go all the way and get the "best" and "latest" regardless of other considerations. Frequently, this divides the organization, with financial management on the one side and everyone else, especially the engineers, on the other.[19] Despite this nonrational method of decision-making, the decisions studied by Bright were eventually successful, and others have since gone further in both the degree and amount of the same kind of automaticity.[20]

Outside of the accountants and the neo-Taylorites, who are discussed below, many people in business seem to be highly skeptical of any and all formulae for evaluating research expenditure, even when basic research is excluded.[21] Nearly everyone agrees that basic research is economically inscrutable and cannot be considered a rational activity within the monocratic intellectual structure that dominates production.

Despite this general skepticism, attempts are often made to apply the controls and evaluations of economic rationality to research activity. To do this it is necessary to estimate the economic value of both inputs and outputs. Even assuming that there is some knowledge about the output (the research objective), the inputs must largely be determined by the researchers themselves, the people who are to be kept within the boundaries of economic rationality by the control techniques. These inputs arise out of the research procedures determined to be necessary by the researchers themselves, rather than by the management, or by some central planning board, or even by the owner of the organization.

Characteristically, the researchers are asked for estimates of the numerical values in the control system, especially estimates of time and cost. Controls thus become rituals that researchers can evade by altering their estimates. Nelson reflects the opinion of many students of this subject when he says that research controls "are almost completely meaningless."[22] A study of weapons development shows that cost estimates are typically off 200 to 300 per cent, and errors in time estimates average two years, or 150 per cent. The errors are greater the earlier in the process they are made and the larger the technical innovation involved. The errors are not systematic; there is a high degree of randomness in the process.[23]

It is my belief that the operation of the control procedure—project control, PERT (Program Evaluation Review Technique), CPM (Critical Path Method), or whatnot—produces the illusion of rationality for some people and therefore a kind of security.[24] The control procedure is fantastically elaborated, as in PERT, and individuals can then evaluate one another according to their correctness or incorrectness in operating the elaborated procedures. If the procedures become complex enough, simply understanding them can be made to appear a respectable professional or subprofessional accomplishment. The flimsy sand upon which the control procedure is ultimately based is obscured by its complexities. A relatively unstructured reality is thus given a pseudostructure.[25]

The inability to bring research within the monocratic structure of economic rationality forces a certain collectivizing of R and D. Otherwise, very little would take place. During the ten years from fiscal year 1953–54, the nation spent about 100 billion dollars on R and D. In the previous ten years it spent only 25 to 30 billion dollars. The nation's 1965 expenditure of 22 billion dollars was over four times as large as that of 1953. In recent years, R and D expenditures have been increasing at about 6 per cent per year. This kind of expenditure is a post-World War II phenomenon.[26] "More money is being spent on research and development now, in this year of 1961, than was spent in the entire period between signing the Declaration of Independence and the end of World War II. And it was so last year, and it was so the year before."[27] In 1934–35, the country spent 200 million dollars, less than 1 per cent of what it spent in 1965.[28]

This great growth is not an argument for the economic rationality of R and D. Most of the money comes from the federal government—65 per cent in 1961–62.[29] The federal percentage of purely industrial R and D has been increasing substantially—from 39 per cent in 1939 to 58 per cent in 1962.[30] This situation reflects federal government interests in regard to the "cold war," including the space competition. Three-fifths of industrial R and D is concentrated in a few industries—aircraft, missiles, electrical equipment, and communications.[31]

The impossibility of incorporating *basic* research within the model of economic rationality—research at the extreme uncertainty end of our activity continuum—is shown by the fact that industry makes only a gesture in that direction. The proportion of industrial R and D expenditure on basic research has remained at about 4 per cent since 1953.[32]

The conflict between productivity and innovativeness is evident in the fact that among the industrial sectors reported on by the National Science Foundation (NSF), the Motor Vehicle and Other Transportation industry has spent the smallest proportion of total R and D funds (7 per cent of total in 1962) and has increased R and D expenditures the least in recent years (only 28 per cent between 1956 and 1962, while Scientific Instruments and Primary Metals, for example, have increased R and D expenditures during the same period by over 200 per cent).[33]

When you compare the twenty-five largest Standard Metropolitan Statistical Areas in terms of the number of scientists registered with the NSF National Register per 1,000 population, you find the Detroit SMSA is at the bottom (Washington, D.C. is at the top).[34] Except for superficial style changes, the automotive industry emphasizes production, not innovation. Perhaps we can get some idea of what constitutes R and D activity in this industry if we contemplate the fact that during a recent three-year period Ford Motor Company had a former missile engineer in charge of a steering wheel research group.

If we divide the SMSAs into those with less than average number of scientists per 1,000 population and those with more, most of the heavy production areas of the country fall into the former group in this descending order: Chicago, Pittsburgh, Cincinnati, Philadelphia, Cleveland, Baltimore, St. Louis, Buffalo, and Detroit.[35] Although industry employs a large and increasing percentage of the scientists of the country (42 per cent in 1962) it makes a relatively small contribution to our fund of knowledge.[36] Under the present conditions of the costliness of research, it seems clear that not much scientific research will go forward without government financing.[37] Increasing the fund of human knowledge is an activity that falls largely outside the economic system.

The current "diabolo" reaction to automation creates an impression of rapid change in industrial processes. "Automation" is simply the latest manifestation of the old and relatively leisurely process of increased mechanization. American industry has not proceeded very far along this route. At present rates of new capital formation it is still sixty to one hundred years behind American technology.[38] Furthermore, technical improvement, greater capital investment, and increasing automation reduce the likelihood of technical innovation in the future,[39] since to discard the old now costs more than it did and may become even more expensive in the future. A smaller and smaller proportion of unit costs is responsive to labor-saving inventions. Automation ties many smaller operations into one large one so that installation, de-bugging, shutdown, and repair costs are much higher. It introduces increasing inflexibility into industrial operations, forcing dependence on an increasingly stable environment in regard to technology, markets, materials, and work force skills.

Since the rational model will not account for organizational innovation, one must look for explanations of a more psychological or nonrational kind.[40] It has been frequently observed that the R and D effort in industry is based, ultimately, on faith. Special efforts are often made to avoid the rational evaluation of R and D.[41] Because it cannot be economically rationalized, research activity is susceptible to budget cuts in times of stress—a fact that tends to create, I should think, a general uneasiness among R and D personnel, rather like that felt by salesmen who are solely dependent on commissions. An uneasy sense of a need to produce commercial results could arise in many people and tend to short circuit success in a kind of self-fulfilling prophecy. A goodly measure of personal security is an important condition for individual creativity.[42] (I believe this is what Albert Einstein must have had in mind when he remarked that no one should earn his living entirely by research.)

The same forces that make R and D especially liable to budget cuts also bring pressure to move the R and D activity toward the production–engineering end of our activity continuum. Where a production department is particularly strong, R and D personnel are liable to wind up spending a large proportion of their time in troubleshooting, testing, customer service, etc.[43]

Optimism becomes important in innovation. We could define it as the excess of subjective probability over objective probability. One condition under which it seems to arise is when the choice lies between winning and breaking even. When the choice lies between losing and breaking even, the two kinds of probabilities are much closer together (the subject is more realistic).[44] This may be why personal security is so important in creativity. Our military R and D contracting procedures have provided incentives toward optimism on both sides, and this fact has been important in the successes achieved. A more businesslike, accurate contracting procedure would save money at the expense of innovation. In fact, productive efficiency and economic rationality are cultural qualities of this country that make innovation difficult.[45] A very general industry practice in budgeting for R and D is to follow some arbitrary rule of thumb, such as a fixed percentage of sales (1 to 1.5 per cent is quite common).

Inability to incorporate R and D within the monocratic intellectual structure of the firm leads to "funny" or "unusual" reasons for getting into this activity ("funny" or "unusual" only because they do not fit within the intellectual structure of economic rationality). Perhaps the board of directors got caught in a research boom and set up research organizations whose use they did not understand.[46] R and D publicity is used to advertise firms, to make them seem progressive and vigorous. An undertone of manipulation and deceit runs through R and D administration. Young graduates are told about "basic research opportunities" in glowing terms, but once they are on the job constant subtle pressures are brought to bear in order to turn them into freely deployable tools of the organization.[47] But this kind of manipulation does not work very well, either, since the grapevine swiftly spreads the information to the universities. It has been estimated that if they had a completely free choice, some 90 per cent of graduating scientists would choose to work in only four or five laboratories in industry.[48]

Industry has found it needs some scientists as a hedge against technological change or for reasons of prestige or for sheer mimicry —but industry severally does not know what to do with them. Recruiting is often governed by the numbers available (take all you can get), not by a rationalized R and D program.[49] In 1962, 30 per cent of the scientists in industry were in administration, 8 per cent were in basic research, 19 per cent were in applied research, 11 per cent were in development or design, 17 per cent were in production and inspection, and 16 per cent were classed as "other."[50] Thus, it seems, only about 38 per cent were working as scientists.

Perhaps the greatest problem is how to reward a growing segment of employees to whom the nature of the work performed is central in their motivation. Organizations have typically held out the hope of promotion into management as the principal incentive for its white-collar employees. Now, however, there are too many employees to be motivated in such fashion, and many of them are not interested anyway. Some firms are beginning to experiment with dual salary-success ladders so that a scientist can "go to the top" without leaving scientific work.[51] Motivating this growing, faceless mass of Ph.D.s (and lesser degrees) will undoubtedly be one of the major personnel administration problems of the future (keeping them up to date will be another). The motivation problem is beginning to show up in terms of growing "shortages" of all kinds of scientists and engineers.[52]

The nonavailability of economic rationality as a decisional device suggests the importance of nonrational factors in innovation—factors such as optimism. These psychological conditions are more likely to exist when the resource picture is fairly lush, when there is slack in the organization. By "slack" I mean uncommitted and unspecified resources of appropriate personnel, finance, material, and motivation; or if such resources have been committed and specified, it has been done in such a way that they are recoverable.[53] A situation in which there is such a slack apparently makes it possible for various psychological variables that are supportive of innovation to operate.

Although this effect of slack has been observed, the mechanism is not well understood. Perhaps the embarrassment of unused resources leads to search for other worthwhile activities; or perhaps a use is found for them simply because they are there. Managerial decision-making on a first-come first-served basis, probably the socially easiest kind for a management, is possible as long as slack exists.

With slack, psychological risk is reduced because the loss of uncommitted resources is psychologically less painful—they have not been discounted in a person's plans and intentions. Such resources are not located on a "means–end map." A gambler with little capital must play a loss control game, using criteria like minimax risk (the best of the worst possibilities) rather than maximizing expectations.[54] I have already mentioned that choosing between winning and breaking even leads to optimistic subjective probabilities. I think that "losing slack" has about the same meaning as "breaking even." The richest farmers are the most innovative, the poorest are the least.[55] One of the most important determinants of the progressiveness of a school system is the wealth of the community, and the same is probably true in other fields, such as public health.[56] Successful organizations are more innovative than unsuccessful ones.[57]

In saying that an organization having a goodly amount of slack can afford to back risky and long-term innovative projects it is important to emphasize that we are dealing with psychological facts. We are not concerned with the rationality or legitimacy of this behavior. Slack at the organizational level is the counterpart of psychological security in the creative process. It makes it easier for management to back innovations. The presence of slack encourages the decentralization of control over resources. Centralized control of resources creates a psychological situation most hostile to innovation. As one goes up the hierarchy of control, the value of the resources controlled increases ominously, while at the same time the knowledge needed to cope with this concentration of responsibility does not increase proportionately, creating a situation of very high risk for the responsible individual. (One would expect governors to be more daring innovators than presidents, and mayors to be more so than both).

Slack can be invested in morale. Thus Raymond Villers, of the Financial Executives Research Foundation, says that you should not fire research personnel when times are stringent, because this will damage the morale of those who remain.[58] Many companies plan their innovations in automation for lush, expanding periods so that none of the current employees will lose their jobs.

The response of an organization to failure or distress, to achievement below the aspiration level, is probably more accurately designated as problem-solving than as innovation. One would expect managerial problem-solving in response to stress to be internally directed and aimed at the despecification and decommitment of resources and the recovery of slack. The more vulnerable, nonoperational, high-risk activities will be discontinued or reduced. Short-run payoffs are badly needed. Such activities as training and research are likely to suffer. If the situation is not too serious, management may be able to recover enough free resources to weather the storm. However, if it has to insist too urgently on the recovery of slack, or if there is insufficient slack to be recovered, the internal health of the organization may be impaired, and after some initial success at recovery, it will spiral downward into catastrophe. Reduction of personnel, raised work standards, reduced supply and maintenance, tightened controls and budgets—all of these measures will free some resources; but in the process other resources of a less tangible nature, such as cooperativeness, good will, loyalty, and ambition, may be used up—offsetting those that are recovered.[59] These considerations point up the importance of organizational slack for organizational survival.

Villers urges industry to determine the R and D budget by allocating all slack to this activity. He says that the R and D budget should equal the difference between expected profit and the desired return on investment—essentially a financial synonym for what we have called "slack."[60] If R and D were dependent upon this source of funds, precious little would occur under conditions of perfect competition. To enlarge the difference between expected returns and a viable rate of returns requires some "market power." According to Schumpeter, innovation involved additional risk costs that had to be passed on, and this required some market control.[61] To him, oligopolistic conditions were necessary for technical, hence economic, progress. The oligopolistic light bulb industry is highly automated (modern, progressive), while the highly competitive shoe industry has achieved little automation.[62] Small marginal industry, apparently uninterested in innovation, pressures government laboratories to ignore basic research and to concentrate on the immediately useful.[63]

Since slack is so important for meeting conditions of stress, it is important to note that the amount of slack is to some extent under management control. Slack is the excess of achievement over aspirations. If the aspiration level can be set below achievement—that is, if satisficing standards (good enough rather than best) can be employed, then the organization will have slack. But even very high achievement will not produce slack if the achievement is not above the level aspired to, since the expected high payoffs will already have been discounted and committed.[64] Even a very productive organization, therefore, may constantly be in a condition of stress, constantly caught up in an emergency problem-solving atmosphere conducive to change but not to innovation. It would seem, therefore, that under highly competitive conditions the managerial posture must be a maximizing one, precluding the accumulation and distribution of slack and greatly depressing the chances of innovation. Furthermore, under free competition competitors will immediately adopt a firm's innovations, preventing a long enough payoff period to encourage much innovation. These considerations, to the extent that they reflect the realities, point up an important difference between individual and organizational creativity and underline the dangers in treating organizations as though they were individuals. Whereas a high aspiration level is important for individual creativity, a low one (below achievement) is important for organizational creativity.

Jewkes and his colleagues say that the statistical evidence does not support either the conclusion that monopoly fosters invention or that competition does.[65] However, our proposition is very different. We say that invention cannot be understood in terms of economic rationality and that this fact underlies the importance of nonrational factors in invention. The term "slack" denotes an objective-subjective condition in which the subjectively set aspiration level has been exceeded by the objective achievement, the excess creating a relaxed, indulgent decision-making situation. Thus, neither monopoly nor competition is the immediately important variable, but rather this subjective-objective situation called "slack." A corollary hypothesis, that slack is more likely to exist under oligopolistic conditions, remains to be tested.

Production takes advantage of existing information, but development (and even more, exploratory or basic research) involves a good deal of new learning. This fact is apparently what Schumpeter had in mind when he said the conditions for efficient allocation of resources were not the same as those for economic growth. R and D activities, therefore, need an open intellectual structuring. Production efficiency can be very costly in R and D.[66]

Reactions to the uncertainties of R and D seem to divide people and their recommendations into two ancient groups, those who blame the uncertainties on men and urge stringent monocratic control, and those who blame the situation and urge a loose pluralistic control. The two ancient groups are the romanticists and the realists.

Frederic Scherer, of the Harvard Weapons Acquisition Project, says that "most variances from original time and cost predictions were man-made rather than caused by an illnatured technology."[67] Thus, he favors "putting tight constraints on the engineers." "Most engineers . . . need to have a set of specific performance and time goals toward which they can work."[68] Typically, he thinks of organization in heroic terms as a social device for exploiting a great or charismatic person—the man at the top who tells everyone else what to do. Thus, he is opposed to multiple approaches to R and D problems because they will overload the topside decision machinery—"the menu of technical alternatives became too rich for a project management group . . . decision makers must indulge in time consuming evaluations before making their choices."[69]

People who think that the uncertainty lies chiefly in the environment stress the development of knowledge as the uncertainty is reduced. In the beginning, all approaches are of equal value because they all generate large amounts of information at relatively small cost, and the unpromising approaches are quickly spotted and abandoned before they become very costly. This new knowledge must be quickly and freely communicated, however, and this rules out all forms of suppression, from competitive suspicion and secrecy to hierarchical insistence on "going through channels."[70]

Programmed or determined behavior prevents the optimum exploitation of this uncertainty. What is needed is flexibility, pluralism, multiple approaches ("duplication and overlapping"). Essentially, this argument is the one used by John Milton and John Stuart Mill to justify freedom. Their argument started with man's basic ignorance, which cautioned him to maintain the freedom of ideas so that truth would have a chance to be discovered and communicated. If we already know the answers, or if truth is a monopoly of some minority, then it is difficult to make a convincing argument for freedom. The need for freedom to exploit uncertainty provides the intellectual underpinning of pluralism. Where there is a minimum of uncertainty, as in production, it is difficult to resist the Weberian monocratic stereotype. Klein believes that the most successful R and D firms (like Pratt and Whitney or Rolls-Royce in jet engines) are "pragmatic," sequential, trial-and-error decision-makers. They are not intellectually structured in monocratic, comprehensive problem-solving terms.[71] Innovation in the aluminum industry increased after Alcoa's monopoly was broken in 1945, apparently because of the opening of additional lines of approach to problems.[72]

A realistic appraisal of any organization that must engage in substantial problem-solving will view it in terms of claimants and negotiation and compromise rather than in monocratic terms of a fixed, single goal set, simultaneous determination of all activities, and constant reappraisals of marginal advantages. One needs only to ask himself what would happen to the United States if all the federal bureaucratic units stopped making demands and sat back and waited for Congress to give them their assignments and budgets.[73] It is even more difficult to think of the President performing this monocratic function. The same considerations apply to any but the most thoroughly programmed, routinized, automated organization. In a pluralistic situation various claimants advertise, propagandize, organize coalitions, and otherwise promote partisan positions with respect to policy, personnel, resource distribution, etc. The organization's decision is largely a result of the claimants' enthusiasm and success in advancing their respective points of view.

We now come to the question: how does this analysis relate to current developments in public administration? In summary form, my answer is that public administration is now undergoing a reaction back to the economy-and-efficiency phase and, even further, to Taylor and the "scientific management" phase. If this is true, and if the foregoing analysis is valid, then public bureaucracy is becoming less innovative.

The social science invasion of administration of the last twenty or thirty years has largely by-passed public administration. Behavioral scientists have been invited to work in business organizations, not public ones, and they have accepted that invitation. Academic public administration is still largely stuck on budgeting, personnel, and operations and methods, just as it was thirty or forty years ago. A quick review of thirteen recent issues of the *Public Administrtion Review* found only 4 articles (out of 91) reporting on original social science research. The National Science Foundation report, *Current Projects of Economic and Social Implications of Science and Technology* (1963), listed no studies by public administration faculties, even in such categories as "Decision Making" or "Administration, Organization, and Management." Out of 27 projects in the latter category, 19 were by business administration faculties. A review of the titles of 256 research projects reported by University Bureaus of Governmental Research for 1965 gave me the strong impression that less than a dozen of them would be of interest to the social science-oriented student of public administration. If public administration education reflects public administration research, it has very little behavioral science content.

A very small proportion of federal government scientists are social scientists (5.8 per cent in 1960), and this percentage seems to be declining. Furthermore, most of these are historians, psychologists, and economists. Scientists in R and D in the federal government constituted only 2.95 per cent of total personnel in 1960, and social scientists in R and D were almost nonexistent—.137 per cent of total personnel. I do not know what proportion of this infinitesimal number of R and D social scientists is doing research on administrative matters, but my impression is that there is almost no social science research of this kind going on. Federal government organizations, for the most part, constitute great unused administrative laboratories.

Many of the people now in charge of administrative matters in the federal government—indeed most of them, I think—received an administrative training that did not reflect the behavioral revolution within the social sciences, either because their training predated this revolution or because, as has been noted above, the revolution tended to by-pass public administration. Some statistics about federal personnel officers serve to illustrate this situation. Almost a third of them are over fifty years of age, only 6 per cent are under twenty-nine. Half of them entered federal service before 1946, only 11 per cent in 1961 or later. More than half of them entered personnel work from another field.[74]

Current administrative practices reflect this lack of modern social science administrative insight. The Weberian monocratic ideal seems to be more strongly implanted than ever before. Decisions are to be made by the man at the top, advised by a few trusted staff aides. The rest of the organization implements them. Thinking, planning, and decision-making are centralized and separated from doing and implementing.

It is surprising to find that the "good soldier" is still the "yes man." Major General W. J. Sutton, head of the Army Reserve at the Pentagon, testified under questioning by a House Armed Services subcommittee that he had not been consulted about merging the Reserve with the National Guard. He was opposed to the merger, he said, but he would support it: "I've been a good soldier for 40 years, and I'm going to continue that record."[75] Those who oppose top policy are often removed (e.g. Generals Curtis LeMay and Jerry Page, and Admiral David Anderson).

This current hierarchical ideal is not just a military one. The Committee for Economic Development recently revived an ancient proposal that the state governor be made the single head of the state bureaucracy. When a Senate Judiciary subcommittee questioned Internal Revenue Service agents as to why they violated the law by breaking and entering and by various forms of wiretapping, they were told by one agent: "Anything that would have been asked, I would have done it." Another one said he would break laws "if my superiors told me to."[76] It is perhaps unfair to point out that this behavior is a regression from the organizational standards espoused at the Nürnberg trials.

The files of the Senate Constitutional Rights subcommittee are bulging with complaints from servicemen and other federal employees to the effect that they have been coerced into buying United States savings bonds and giving to charities. As of September 30, 1966, for example, some 1,427,602 poorly paid servicemen were enrolled in the payroll savings plan! The Monthly Staff Report to the Senate Subcommittee on Constitutional Rights is rich in other illustrations of the Weberian bureaucratic stereotype. Some examples will be interesting:

> In one case, a Vandenberg Air Force Base memorandum ordered employees to hold personal contacts between employees and the Personnel Office to a minimum, and prohibited phone calls to that office without supervisory permission, or personal visits without written permission.
> In the other memorandum a Federal Aviation Agency supervisor informed his division of possible reprisals and effects on their standing in the agency if they wrote to their Congressmen or exercised rights under the agency procedures to file a grievance complaint. (2/1/67)

[These cases are good illustrations of the difficulty of making an administrative appeals system work in a monocratic institution where conflict cannot be legitimatized.]

> Under one Department's regulations, employees are requested to participate in specific community activities promoting local and federal antipoverty, beautification, and equal employment programs; they are told to lobby in local city councils for fair housing ordinances, to go out and make speeches on any number of subjects, to supply flower and grass seed for beautification projects, and to paint other people's houses . . . employees were being

informed that failure to participate would indicate an uncooperative attitude and would be reflected in their efficiency records. (11/1/66)

A captain in Massachusetts stated that junior officers are expected to display their leadership ability by getting 100% participation from their units. (4/5/67)

The weakness or absence of the social science voice in public administration has left the ground clear for a naïve invasion of the subject by nonsocial—perhaps even antisocial—science. The physical sciences provide models that unify the fields, define the important problems and the criteria for their "solution," and provide examples of "good" scientific practice.[77] With such an intellectual control apparatus at their disposal, scientists experience a high level of "success" in solving problems. Most of them do not realize how relatively easy it has been made for them. They are presented only with solvable problems, and their "solutions" are solutions only with reference to the single set of conventions of the field. There is no pluralistic babble of voices to challenge their "successes" or—perish the thought—to impugn their motives. As C. P. Snow said, they become optimists, imbued with the idea of progress and the related notion that science can also solve social problems. They are immunized from the laity both by tradition and by their textbook educations.[78] It is from this background that the new "scientific management" has risen.

In 1916, Henry Gantt wrote: "The substitution of fact for opinion is the basis of modern industrial progress, and the rate of this progress is controlled by the extent to which the methods of scientific investigation supplant the debating society methods in determining a basis for action."[79] In 1935, in view of the approaching crisis, Great Britain established the Committee for the Scientific Study of Air Defense (the so-called Tizzard Committee). C. P. Snow reports that this committee had to teach the military a lesson. "The lesson to the military was that you cannot run wars on gusts

INNOVATION AND ORGANIZATIONAL DECISION-MAKING   53

of emotion. You have to think scientifically about your own operations. This was the start of operational research."[80]

Although the earlier "scientific management" of Frederick Taylor and his followers was kept in check by unions and even congressional investigations, and by the groundswell of social scientific research following the Hawthorne investigations,[81] today the federal government is caught up in a resurgence of "efficiency and economy," of so-called scientific management, of neo-Taylorism. As the Bureau of the Budget has recently stated: "The subject of productivity measurement in government received added significance with the new emphasis on efficient utilization of resources in the Government. Both the late President Kennedy and President Johnson stressed the need for productivity improvements in each agency and the need for proper measurements in this area."[82]

President Johnson has ordered all federal agencies to switch to a new "planning-programming-budgeting" (PPB) system of control. "The new . . . method is an extension of the 'cost effectiveness' approach [that] Defense Secretary McNamara has been using at the Pentagon. This method will require each agency to spell out specific goals and objectives, set forth different methods of achieving them, attempt to measure exactly what results will be achieved for each dollar spent under each method—and [to] project all of this not just for one year but for five years."[83]

The people who have picked up this new opportunity for acquiring administrative power are not social scientists. They are applied mathematicians—econometricians, operations researchers, computer programmers, decision theorists—to whom I will refer as "econologicians." They are *econo*logical rather than *socio*logical in orientation. The scientific study of administration, as they conceive it, *must* end as the scientific administration of things, including study.

Robert Boguslaw calls this new nonsocial science administrative power elite the "New Utopians," and the designation is apt, since the econologicians habitually make the same mistakes as all past utopians have made. Specifically, they vastly underestimate the complexity of the units with which they deal (individuals and social systems) and they fail to take adequate account of emergent situations.[84] They repeat the mistakes of Taylor so completely that it seems quite accurate to refer to their viewpoint as "neo-Taylorism." By way of illustration, consider the underlying assumptions of the new PBB system of control.[85] These assumptions are:

1. The organization can be disregarded; only the means–end logical analysis is relevant.
2. The techniques of mathematicians and accountants, not those of social scientists, are the basic managerial tools.
3. Human motivation is not problematic; it can be assumed.
4. The only legitimate concern of administration is production values (or else, all kinds of organizational outputs vary together in the same direction—a concept of "the one best way").
5. Rationality involves comprehensive analysis of a problem—"getting all of the facts"—and a balanced, coordinated movement toward a single preset goal group or preference-ordering.
6. Social and cultural conditioning are either nonexistent or irrelevant.
7. A single preference-ordering for all social needs is possible (or perhaps only desirable).
8. It is worthwhile to achieve the *form* of rationality even though the content is not rational, that is, to decide by defendable formulae no matter how arbitrary the values assigned to the variables may be.
9. Organizational flexibility—the absence of an organizational claim—is assumed. (This implication is surely *not* realized by the top managements that support this approach!)[86]
10. PPB rejects acceptance, consensus, or other sociopolitical criteria for achievement. People do not know. You must use

machines—logic, mathematical manipulations. (The new Taylorism is also a new Platonism.)

Perhaps unconsciously, this new power elite uses a characteristic power-acquiring technique. It involves drawing attention to certain kinds of deficiencies in present management decision-making that should make management dependent upon the new elite—the econologicians. A hypothetical case in a PERT instruction manual says: "Having only the vaguest conception of what is involved in the installation of a large scale data processor, Ferris [vice president for administration] calls in young Wiley, a staff procedures analyst."[87] Or note the contemptuous tone of the following: "The board, after piercing questioning and solemn deliberation authorized Wiley to begin the project."[88]

A favorite technique of this group is to examine decisions, discover the underlying rules of thumb, and then look for the assumptions under which the rules of thumb used would be optimal. Thus, if the decision-maker is reluctant to accept the assumptions, "then he ought to reexamine very carefully his acceptance of the rule of thumb. It is possible that he is being seriously inconsistent."[89] We are threatened with serious consequences if we do not change our ways: "But it is very clear that we are facing a period in which reliance on these [rules of thumb] to the exclusion of other ways of managing will leave us in serious trouble."[90]

The logical effect of these criticisms, if they are accepted, is to make management dependent on the neo-Taylorites and, thus, to create a new power group in the organization. "The 'management system design function' is one which is being established in many organizations and serves as a buffer between management and the computer lab. The specialists in this organization know the needs of management and are also able to communicate effectively with the computer specialists."[91]

The neo-Taylorites belittle present decision-making, but they have no new theory telling them what additional facts are needed beyond the production data of the present accounting system. They assume that past failings have been in the process of decision ("not enough math"). In fact, however, most failings have been caused by insufficient data and inadequate predictive theory, especially sociopsychological data and theory about variables that intervene between management decisional inputs and production outputs[92] —the very kinds of data in which utopians (systems designers) have always been deficient. Pumping imaginary data into equations in no way solves the problem.[93]

This lack of new substantive answers is concealed by means of complexly elaborated procedures. The neo-Taylorites set up self-serving rules that assure their being able to reach determinate solutions. That is, they solve what problems they can, not the problems that most need to be solved. This makes them "look good," like the physical scientists. They optimize a set of conventions, not subjective utility. Perceived welfare may very well decline; demonstrable welfare increases.

The growing influence of this new elite group is ushering in an era of "scientism," which may be defined as the belief that "science" can solve social problems. Thus, Dr. Glenn Seaborg, a chemist, assures us that there is a "technological fix" for *every* problem. Representative Bradford Morse of Massachusetts wants a "national commission" to consider ways of applying modern management techniques to the solution of public problems. He and Senator Gaylord Nelson want to apply something called "the system approach" (taken over, presumably, from aerospace technology) to public problems. The Department of Justice has hired the Illinois Institute of Technology to "solve" the problem of crime.

This naïve belief in "science" and associated econological techniques has apparently dominated recent thinking in the Pentagon. For example, Secretary Robert McNamara's approach to the antiballistic-missile (ABM) problem was based on this kind of simplistic, mechanistic thinking. It was reported that he said a Chinese missile attack on the United States by the late 1970's would kill only 10,000,000 people even if nothing were done to provide a special defense against such an attack. By contrast, a Soviet attack would kill 90,000,000 people. If we installed an "austere" three and a half billion dollar ABM system, the Secretary promised, we could reduce the number of deaths to less than 1,000,000 people. Further, with "modest additional outlays" damage could be kept to "low levels" beyond 1985.[94]

The danger in "scientism," as the above example illustrates, is that such calculations are actually taken seriously, both by decision-makers and by the people they must impress. It should be remembered that, although the form of the calculation is undoubtedly impeccable, the values assigned to many of the variables are almost completely arbitrary. The danger lies in the fact that the decision-makers may come to forget this—assuming the technicians took them into their confidence—and the fact that others are not in a position to refute the conclusion. The mathematical form of presentation tends to intimidate would-be critics into silence, and the resulting decision has such utility for the decision-maker that he, too, forgets the flimsy grounds upon which it is based. For some, it seems, the illusion of certainty is just as comfortable as the real thing.

I must be blunt: science cannot solve social problems. Suppose, for example, that we ask medicine to solve the problem of race prejudice. As a medical problem the "solution" might turn out to be some drug. However, the *social* problem would still remain. Should "we"—alternately, "they"—put the drug in everyone's drinking water? Does the federal government have jurisdiction to do this? States? Counties? Any "scientific" (chemical, physical, medical, biological, etc.) "solution" to a social problem is almost certain to sound ludicrous. The solution of a social problem is properly described with such words as "compromise," "consensus," "majority," "negotiation," "bargaining," "coercion," etc. If the "solution" cannot be described in such terms, then it is not the solution of a social problem.

Econologicians are "systems" thinkers. As Boguslaw says, they are usually confounded by emergent phenomena. It is reported, for example, that Secretary McNamara tried to analyze and evaluate the Vietnam war by means of "cost-effectiveness" methods. Ever so typically, he was highly optimistic in the early days of the war. Cannot scientific thinking solve any problem? "But unforeseen forces kept breaking through to drown his expectations: the Buddhists, the collapse or overthrow of this Saigon leader or that, the inability of various strategic calculations to produce decisive results. 'He changed from a mechanical optimist to a mechanical pessimist,' is the way one who knows him intimately has put it."[95]

The reemergence of Taylorism and the "efficiency and economy" orientation in administration is facilitated by the absence of social science in administration and by the rapid growth of a whole industry dedicated to control-oriented management and attempts at comprehensive economic rationality.[96] The approach of the applied mathematicians is much closer to the old pre-Hawthorne tradition in management. In truth, the neo-Taylorites are simply the older "procedures writers" with mathematics added. Management seems to have few defenses against this onslaught.

Federal administration appears to have succumbed to neo-Taylorism. Econologicians are "playing store" with the federal government. Spreading from the Rand Corporation to the Department of Defense, neo-Taylorism has since moved into the Budget Bureau and throughout the Administration of President Johnson. I have already mentioned that the President has ordered all agency budgets to be run on a "cost-effectiveness" basis. As this is written there are about ninety federal employees being trained in these procedures at various universities, and plans call for about one hundred and fifty more to be trained in 1967–68. Thus, at first

INNOVATION AND ORGANIZATIONAL DECISION-MAKING

glance, there seems to be little to stop us from a mad dash into Orwellian mechanicism. There are, however, certain counteracting forces.

Social systems, like biological systems, have ways of protecting themselves from various threats. One way is to disobey or otherwise evade threatening rules or systems. For example, the many inequities in the income tax laws are tolerable, in the degree that they are, only because people can engage in minor cheating and evasion. If the income tax collection system were mechanized to the point of being completely effective, the laws would have to be completely overhauled or else the system would break down for lack of citizen cooperation. The weakness of machines is that they always obey (although you could, of course, have a disobedience *rule,* which a machine would then always obey),[97] and they cannot, therefore, accommodate emergent, or innovative, situations and conditions. A designed social system, or utopia, is a machine and suffers from the machine's defects. Such a social system may be saved, however, by *unprogrammed* disobedience by human beings.

Given the necessity of claimancy and partisanship (top-down comprehensiveness is a managerial delusion), procedures that ignore the organizations which structure this claimancy, or that attempt to substitute a comprehensive, rational "logic" for it, must either work toward organizational breakdown or be evaded by various stratagems. I have no doubt that one of the most intensely pursued activities in Washington today is finding ways to evade PPB, ways to manipulate it so that "you can live with it." We can only speculate about this, however, because in the humorless neo-Taylorite world of modern management, control systems are almost invariably studied according to the logic of the controller's purpose rather than according to their latent effects—the adaptations of those being "controlled."[98]

It has been argued, however, that devices like PPB are more important for their latent functions than they are for their manifest functions. Such devices may not tell us whether we are getting "our money's worth" for every "defense dollar," or whether we should allocate more money to education. Even so, it is said, they do encourage cost consciousness and they do draw attention to the interrelations between activities.[99] Economy and consistency are undoubtedly proper matters of concern, and they should be represented in decisional processes, but surely there must be some way to promote these values without sacrificing all spontaneity, originality, freshness of approach. The therapy of PPB is too drastic by far. Economy and consistency should be represented, but they should not dominate. They are not of themselves the final goals of this or any other society.

The monocratic, comprehensive, top-down intellectual structure of latter day Taylorism may be compatible with activities at the certainty end of our activities continuum,[100] but it becomes more and more dysfunctional as we move toward the uncertainty end—toward innovation. Whatever one's opinion about management in the Department of Defense, currently the mecca of neo-Taylorism, the Department still has plenty of critics with regard to new weapons R and D.[101] Though we still lack criteria by which to measure innovation with any precision, indirect evidence and analysis point strongly to the conclusion that the United States is passing through an unusually noninnovative period in public administration. Government is seemingly stuck dead center on many important matters of basic policy, foreign and domestic, and although much day-by-day administrative puzzle-solving takes place, nothing very creative is going forward. Far from being innovative, public administration is in a period of reaction.[102]

IV

A Program of Research on Innovative Organization

WHAT is most obviously lacking in present day organization theory is a unifying model of this particular universe—a model to define the important problems, to provide methods and standards for their solution, and to illustrate good scientific practice.[1] Without such a model the field is individualistic, chaotic, and non-additive.

If we accept the *Handbook of Organizations*[2] as a fair sample of the field, the utter confusion becomes painfully obvious. Also obvious, and very surprising, is the fact that organization theory has been little influenced by social science. In this field, indeed, social systems are being studied by people who do not accept the reality of social systems. The econologicians say that organization theory "deduces propositions about the behavior of an organization from assumptions about the organization's members and its environment."[3] Decision-making theory is as applicable to General Westmoreland as to General Motors and quite generally confuses justification with explanation. Psychologists study the relations between groups and the individuals composing them as though roles and norms did not exist. Foremen are urged to be "employee-minded" in complete disregard of the fact that if they behaved that way they would eventually be fired.

In this "interdisciplinary" field, most disciplines are self-centered and proceed in complete ignorance of the relevant work that has been done and is being done in other disciplines. In the *Handbook*, for example, the sociologist Stanley Udy reports on "a general survey of recently published empirical literature" concerning "The Comparative Analysis of Organizations" without once referring to Fred Riggs and the growing field of Comparative Administration in Political Science.

In some cases the fragmentation goes beyond the discipline to the individual level. Thus, Amatai Etzioni is not even bound by the traditions and consensus of sociology. He regards an organization as something held together by power and threats. "Most organizations most of the time cannot rely on most of their participants to carry out their assignments voluntarily, to have internalized their obligations."[4] In this anarchy, research findings are not related to important theoretical problems, and the research going on is more likely to be mutually destructive than it is mutually reinforcing and additive.

During the past year or so, I have been taking a hard look at the theory and research in the field of organizational innovation. I have found that there has been surprisingly little organizational research of any truly meaningful kind. There is a plethora of research on individuals and small groups that may be applicable to organizational problems in some way. There is less, but still a sizable amount, of research on individuals and small groups *in* organizations, particularly those at or near the bottom of the power pyramid. But we have very little research on organizations as such. We do not even know how to measure them so that we can describe them accurately. We have subjects and courses such as "organizational pschology," but they deal with the psychology of individuals or groups *within* organizations.

An "organization theory," to be of real utility, must be able to deal with organizations, as such, in terms of measurements. In this chapter I want to suggest an approach to organization measurement, the kind of research needed to do this, and the kinds of research that would then become feasible and necessary for the development of an "organization science."

The first requirement is a concern or interest—a reason for studying organizations in the first place. Some of the difficulty heretofore has been the tacit assumption by some people that one could simply study the organization itself, as a sort of *Ding an sich,* from a detached viewpoint of cosmic curiosity. Most students, however, have studied human organization from the viewpoint of a concern rarely made explicit. Most often this has been a concern with production—producing more and more of the same thing at less and less cost. Sometimes it has been the empty "survival" of an uncritical functionalism. My own interest, or concern, is in innovation. As I have sought to show in previous chapters, there is some research and theory about innovation within organizations. Using this concern, and the small body of theory about innovative organizations and related subjects such as individual creativity, I am able to suggest a series of significant organizational measurements—significant from the standpoint of innovation.

Since our immediate objective is to get solid, objective measurements of organizations (units, branches, departments, or whole companies, depending upon the problem), we should avoid qualities that depend largely or entirely upon the personality of the person who happens at the moment to be occupying a determinate authority role. He may move on tomorrow.

Our measurements will result in a series of indices by which we can specify an organizational unit in quantitative, objective terms that can be compared with others. It should be noted that these indices will be meaningful descriptions only in terms of an interest in innovativeness and a body of research and theory that relates these measurements to innovativeness. Eventually, with the development of these measurements, research will relate level of innovativeness to other output variables and then the indices will acquire this additional significance. The indices can be combined into a single composite index of innovativeness.

In the past, as I said above, interest in organizations was usually limited to production-type values (output per x, unit costs, etc.) and measurements and resulting indices were only such as related to this interest. It now begins to appear that theory and research evidence relating these indices to this interest were quite incomplete and superficial; it was never possible to *describe* the productive organization in meaningful terms, but only to point to it. ("That organization is productive, but we do not know quite *why*.")

With some experience, it should be possible to take other measurements in terms of other interests (productiveness, employee satisfaction, etc.) and thus derive additional indices. It is almost certain, in my opinion, that some of these interests and concerns—innovativeness and productiveness, among them—will be mutually incompatible. Consequently, administrative objectives will have to be set in terms of trade-offs—so much productiveness for so much innovativeness, for example. From the standpoint of such an objective, the ideal organization would not be the one highest on all or any of these incompatible indices, but it would have some kind of optimum mix, depending upon the mix of these incompatible interests or objectives. The point is that the ideal organization could be described or specified in terms of a whole set of indices, each based on the results of solid measurements. The importance of this accomplishment both for management and for organization research can hardly be exaggerated.

People within organizations are related to one another in several ways—by authority, technically, by status rank, and, from the operation of these three, by social relations. As the initial unit of measurement, I suggest that we use the authority relationship or grouping of a superior and his subordinates. Elementary authority units can then be combined into larger, more comprehensive groupings (units to sections to branches, etc.), according to the problem to be solved. I suggest we start with the authority grouping as the unit for organizational measurement because the authority subsystem is usually also a goal or mission assignment subsystem. It normally overlaps with other subsystems more than any other grouping would. Consequently, organizations so defined are usually more real than any other in terms of interaction, formal and informal structure, and the solidity of traditions. However, important exceptions are to be found in project or committee organization, and occasionally in *de facto* groups created by the technical subsystem (e.g., interdisciplinary research collaboration). From the existing body of theory on organization innovation it is reasonable to hypothesize that certain kinds of project organization would get the highest possible score on the composite innovation index.

In previous chapters I have presented the theory of organizational innovation upon which the measurements are based—the theory that relates the measurements to organizational innovativeness. This theory is, of course, rudimentary and deficient in many respects. An operational definition of innovation has not yet been agreed upon, and measurement of this output variable is in a chaotic state. Much of the best research has dealt with the problems of scientists and engineers in research, development, or engineering in bureaucratic organizations rather than the specific problem of organizational innovation. In fact, I have found very little of this kind of research, and that was almost always limited to technical innovation. Nevertheless, from the theory presented thus far a good start on an organizational research program can be made.

Before presenting the research program it would be well to deal briefly with the problem of measuring innovation itself.

How can you tell which of two organizations is the most innovative? The National Science Foundation is exploring the possibility of using: (1) items relating to publications, such as citations, (2) data on patents, and (3) data on the potential economic value of innovations.[5] In addition to these measures, economists have tried to use the amount of R and D expenditures. None of these approaches works very well. The expression "R and D expenditures" does not have an unequivocal meaning, and since research and development is currently a prestigious fad, the figures reported to NSF are not completely trustworthy.[6] The 1954 Revenue Act exempts R and D expenditures from taxation—a further inflating influence. Only 4 per cent of R and D expenditures is for basic research, and we do not know what proportion is for inventive-type activities. Most of it—68 per cent in 1961–62—is for development.[7]

The purchasing power of the R and D dollar is most likely deteriorating more swiftly than that of the dollar in general. There has been a rapid increase in researchers' salaries, the biggest expense item, owing to an increasingly cut-throat competition for personnel that is caused, in part, by a government grant-and-contract policy that allows higher salaries for new personnel than for old. This policy encourages rapid turnover and declining efficiency (since it generally takes a couple of years on the job for a researcher to begin to "pay off"). As more people are brought into R and D activities, the average level of competence should become lower. Growing size and age of research establishments most likely are paced by growing bureaucratization.

All of these things lead one to expect diminishing returns—perhaps rapidly diminishing returns—from R and D expenditure in the future.[8] Between 1880 and 1955, input in the form of trained personnel (engineers, chemists, physicists, professional R and D staff) increased 226-fold while output in the form of books and articles abstracted yearly in such indexes as *Chemical Abstracts* and *Engineering Index* increased only 86-fold.[9] Moreover, since we lack much in the way of an adequate theory of organizational innovation, we have little understanding of what comes between sheer dollar inputs and outputs in this field. We can spend a great deal of money for innovation and still not get it. A National Science Foundation study of 1953–54 showed that R and D expenditure will not predict the number of patent applications.[10]

Patent data are also unsatisfactory for purposes of measuring innovation. The patent rate in this country remained constant (at about 1 per 2,700 population) for two generations after 1885, whereas in terms of increased inputs one would have expected a severalfold increase over that period of time. Much the same thing happened elsewhere.[11] Furthermore, the rate has been declining swiftly in recent years. In 1956–60 the rate was 1 patent per 3,471 persons, and in 1964 it was 1 patent per 3,830 persons.[12]

Any activity that involves uncertainty will generate knowledge, which may be defined as the reduction of uncertainty. Since knowledge is communicated principally by the written word, the number of publications should be a measure of innovation. In recent years, however, the use of publications as a status- and income-producing technique, especially in the universities, has compromised the publication as a communication device. A great many, and perhaps most, publications are of little or no value. The differences in quality are so great that simply counting publications, as such, is like adding apples and donkeys.

Furthermore, administrative pressure to publish in the universities results in enormous duplication. If journal articles are used as the measure of innovative output, we will have to conclude that almost no innovation comes from industry. Of 2,340 articles in *The Physical Review* during 1956 and 1957, for example, only 10 per cent came from industry. There are reasons that we should expect relatively little innovation from industry, as we have seen, but surely such a small proportion indicates some flaws in the measuring device. The use of citation counts has not solved the problem of measurement, as they have proved to be very unreliable in predicting what articles experts will pick as truly valuable.[13]

Case studies in the form of lists of important inventions are of even less value for measurement purposes. Using them it is impossible to show innovative progress at all.[14] One soon becomes disenchanted with the use of case illustrations or counts in this field. It is so easy to find cases to "prove" any proposition whatever.

The study of diffusion or speed of adoption has been a promising approach. In areas where new products or procedures seem obviously to be in the interests of prospective users, it is informative to study the varying speeds of adoption. Such studies have been fruitful in agriculture and medicine and they are beginning to be used in other areas such as public health and education.

Diffusion studies are actually studies of technological change. They have not yet been sufficiently integrated theoretically to be of general use. Lawrence Mohr, of the University of Michigan Public Health research project, after reviewing these studies, concludes: "Innovation . . . has been linked to size, wealth, ideology, motivation, experience, competence, professionalism, nonprofessionalism, social status, opinion leadership, and still other variables."[15]

A measuring device that is just beginning to be used is the rating of organizations for innovativeness by knowledgeable outsiders, insiders, or both. Ratings have been used substantially to study production, and almost everything else, but they have not been widely used in the study of organizational innovation.

Unable to find workable objective measures of innovation in the literature, Chris Argyris has substituted measurement of individual psychological properties as observed in interpersonal relations within organizations—willingness to acknowledge one's own ideas and feelings, openness to new ones, and willingness to risk one's self-image with new ideas and feelings. He finds these qualities appallingly absent from the organizations he studied, one of which was an industrial research lab.[16]

At this writing, it seems that research in organizational innovation will have to make pragmatic use of whatever measurements are available in the context of the specific research project. In time, perhaps, strong relationships will be discovered, and we will gradually get a better idea of what we mean by an innovative organization. In the interim, however, we should get on with the measurement of those organizational qualities which theory tells us are related to innovativeness.

INDICES RELATED TO PROFESSIONALISM[17]

A professional has a depth of knowledge that allows him to work at the perimeters of a field, where innovations occur. He is also less likely to be strongly oriented toward the organizational values of money, status, and power, all of which, as we have seen, tend to promote conformity rather than innovation. He is also very likely to have internalized certain professional values that seem to be related in some fashion to innovativeness. These are: (1) autonomy in work, both as to means and ends, (2) a belief in professional growth as the measure of success, (3) an acceptance of peer evaluation, rather than the opinion of a "superior," as the standard of personal worth, (4) an assignment of the highest value to activities that develop new knowledge (pure research over applied research, etc.). Some professionals are more strongly identified with these values than others (e.g., scientists more than engineers, but engineers probably more than lawyers). Professionals with more education are identified with these values more strongly than those with less, and young ones more strongly than their elders. Further comparative studies of the professions are needed to complete our information about these professional differences.

Set forth below are some suggested measurements or indices

that follow from these propositions. The procedure is two-fold: (1) get a score for the unit under consideration, and then (2) divide this score by the score for the organization as a whole (or branch or division or department, etc., depending upon the purposes to be served). In other words, all of the measurements are relative or comparative, the standard being another organization (just as the standard of a weight is *another* weight, the standard of a length is *another* length, etc.).

$$\frac{\%\text{ of scientists in unit}}{\%\text{ of scientists in whole organization}}$$

$$\frac{\%\text{ Ph.D.'s in unit}}{\%\text{ Ph.D.'s in whole organization}}$$

$$\frac{\%\text{ professionals in unit}}{\%\text{ professionals in whole organization}}$$

$$\frac{\text{unit }\%\text{ of professionals who are engineers}}{\text{organization }\%\text{ of professionals who are engineers}}$$

This last measurement needs a great deal of work. We know that professions vary widely in the extent to which they indoctrinate their members with a strong professional identification. We know, further, that engineers are probably near the bottom of the list, but we do not have this information for many professional groups, and hence we do not have an adequate ranking of professional groups in this respect. Moreover, if the implementation of new ideas is included in a definition of innovation, as I think it must be, then the innovative implications of these professional variations is not clear. Since we are unable, at the moment, to rank professions according to their ability to instill strong professional identification (hence, according to their commitment to professional values and incentives), it might be better, if feasible, to measure these identifications and orientations by direct survey. But even if a certain amount of survey work is now necessary, our objective should be to fill out our knowledge of organizational behavior so that our dependence upon this kind of expensive and precarious measurement is reduced. It certainly should not be necessary to take the same survey measurements over and over again in different organizations, as though the survey results were completely ephemeral and localized.

A method that seems to have produced good results is to have the professional members of a unit rank a few sources of rewards as to their importance, such as:

1. "opportunity to contribute to scientific knowledge";
2. "respect of fellow professionals because of my achievements";
3. "association with other professionals of recognized ability";
4. "membership in a company producing reputable goods and essential service."

This procedure will apparently make it possible to classify professionals according to their respective orientations—organizational, professional, or mixed. According to William Kornhauser (*Scientists in Industry*), those with mixed orientations are likely to be interested in applications, and hence they may be important in getting ideas adopted and as intermediaries between creative innovators and organization men. I omit the nonprofessionals from such measurements because they are either organizationally oriented or alienated, and in either case their lack of occupational preparation means that important innovations cannot be expected from them—except, of course, in the case of an unusual individual. (The unusual individual does not enter into these measurements and this research. He is a problem for psychology, not organizational science.)

A final measurement with regard to professionalism is the following:

$$\frac{\% \text{ professionals-turned-supervisors in unit}}{\% \text{ professionals-turned-supervisors in whole organization}}$$

As we shall see below, this measurement is important in protecting boundary relations between innovative areas and noninnovative environmental elements, and consequently in protecting the former.

The Integrative–Aggregative Index

It is hypothesized that, for a number of reasons, an interdisciplinary organizational unit, composed of many different professional or subprofessional roles, will be much more innovative than an aggregation of individuals all performing the same nonprofessional (desk-class) work, like a section of auditors. The reasons relate to the greater depth and variety of idea inputs, greater interpersonal communication and stimulation, stronger professional orientations (e.g., the search for novelty is more likely to be personally rewarding), and similar qualities, in the integrative than in the aggregative unit. A simple measurement will do. Excluding supervisory and clerical positions, divide the number of different job titles or job descriptions of different series (i.e., junior chemist, chemist, and senior chemist constitute a single job description) by the total number of positions. This measurement can be applied to a unit, a section, a branch, etc. The higher the percentage, the greater the degree of integrativeness. The unit that has only one nonclerical, nonsupervisory position, so that it would get 100 per cent on this index, is to be regarded as an aggregative unit; there is no technical interaction between operatives.

Parochialism–Cosmopolitanism Index

The extent to which an organizational unit is obsessed with its own affairs, rejecting or neglecting the affairs of other units, is indirectly related to organizational innovation in a number of ways. It greatly affects the quantity, variety, and quality of idea inputs. It affects the extent of multigroup membership (potential alliances), which in turn affects both input and support both for invention and for implementation. Parochialism is resistant to ideas from "outside" and also promotes waste of resources by duplication of effort and refusal to cooperate in regard to sharing information, planning, and implementing plans. However, by way of caution, it should be noted that an innovative unit within a noninnovative organization may survive by adaptations that appear to the observer doing the measurement as a high level of parochialism. Boundary-protecting adaptations are discussed below.

A method for constructing this index is as follows. First, by conversations with knowledgeable persons in the organization, develop a small list of organization-wide, newsworthy events of the past few years. The members of the unit to be measured are then polled as to their knowledge of these events and an average unit score derived. The parochialism-cosmopolitanism index is then the average score of the unit divided by the average score of the whole organization. This measurement can be made more precise by dividing the unit score by the average score of a class of comparable units—say all units with over 75 per cent of nonclerical personnel, etc.

A related measurement, possible when the organization under study has a suggestion system, is the percentage of suggestions accepted by a unit in the organization divided by the percentage accepted by the organization as a whole.

Obsolescence Index

With the rapid increase of knowledge of all kinds, maintaining a degree of professional depth and currency sufficient to promote innovation requires continuing education of potentially innovative personnel. Much of this responsibility will be shouldered by individuals, but an organizational measurement of this factor is becoming very relevant. It is hypothesized that an innovative organization will be well above average in encouraging continuing education. A simple measure would be to find the total annual average training man-hours over a recent period and to divide this figure by the total annual man-hours of the unit. "Training man-hours" should include all time spent in class, in-house or otherwise, whether subsidized in some way or not; man-hours spent attending professional meetings; and time spent in informational seminars held by the organization. The measurement would then be the unit's percentage of training man-hours divided by the percentage for the whole organization.

The trouble with this kind of measurement, and that of close supervision mentioned below, is that it could reflect the chance appointment of a training-oriented supervisor. More generally, however, it will reflect the degree of professionalization. It is the sort of measurement that may well become superfluous when we know more about how the various measurements are related.

The Index of "Purposiveness"

This index and its theoretical significance need much more work. Managerially oriented thinkers stress the importance of purposiveness, but I think this is because managerial theory has been dominated up until now by an implicit or explicit interest in production values alone. Purposiveness undoubtedly plays an important role in production. I would hypothesize that innovativeness, however, is related to goal vagueness and ambiguity—not to the absence of goals but to their vagueness. I would hypothesize, further, that a high level of purposiveness leads to centralized control and programmed behavior and in any case rules out value innovation. In an age of information affluence value innovation becomes very important—finding new uses to which our information can be put.

Purposiveness could be measured in the following way. From delegation documents and discussions with knowledgeables, work out a list of goal statements for the organization unit to be measured. Then by polling the members, measure the extent of agreement on goals within the unit. This measurement will indicate the extent to which there is agreement concerning the unit's goals and is sufficient of itself. If the hypothesis proposed above is correct, the measure will nicely mark off the innovative areas from the rest of the organization.[18]

The Index of "Programmedness"

This index, which measures the extent of perceived routinization, is obviously closely related to the index of purposiveness. If a person's activities are completely programmed, there is no room for innovation—except on the part of the programmer. Although more objective data would be preferable, those obtained by means of a questionnaire are perhaps the best data available to us at present. The nonclerical, nonsupervisory members of each unit being studied can be asked to choose one from among the following statements:

1. For most of our actions there are clear rules or procedures prescribed by the organization.
2. For many of our actions there are clear rules or procedures prescribed by the organization.

3. For some of our actions there are clear rules or procedures prescribed by the organization.
4. For very few of our actions there are clear rules or procedures prescribed by the organization.

The larger the percentage of employees choosing items 1 and 2 combined, the greater the perceived routinization. The larger the combined percentage for items 3 and 4, the larger the perceived discretion. These percentages can be compared with percentages for the whole organization.

Somewhat more objective measures of autonomy are possible, but their validity has yet to be established. For example, the average number of visits of a unit's supervisor(s) with nonclerical personnel (once a day, once a week, etc.) can be compared with the organization-wide average. Or the percentage of the unit's conference man-hours (of total man-hours) can be compared with the percentage for the organization as a whole. This measurement assumes that most conferences are held largely for the purpose of solving problems, and that the more people there are who are involved in conferences, the more there are who are involved in problem-solving. It is quite possible, however, that conferences serve other, more obscure needs (such as the reassertion of group solidarity and integration, or ritual displays of loyalty and subordination to an insecure superior) and that total conference time may be only randomly related to the amount of problem-solving activity (as opposed to programmed or routinized activity). I am inclined to think, however, that conference activity usually reflects problem-solving (indeterminacy, nonprogrammedness) and that this measurement is, therefore, valuable.

Index of Communication Openness

Since, according to the theory, innovation is related to richness and variety of idea input, and since implementation of results is related to informal relations with potential users of new ideas, communication is a crucial variable. A number of measurements are potentially relevant here. Nonclerical personnel of the unit under study could be asked to estimate the percentage of their time that is spent reading informational materials (reports, articles, etc.). The average for the unit could then be compared with the average for the whole organization. Nonclerical personnel could be asked the number of communications they had had the previous day from or with persons outside their unit (i.e., with persons other than their superior and fellow subordinates). A "communication" would be defined as a face-to-face or telephone conversation, a committee meeting, or an interoffice memorandum (any written communication other than a formal regulation or instruction). The average number of such contacts over a period of time would be more useful, of course, but these data would probably not be obtainable. If they were, however, then the average number of communications per person for the unit could be compared with the average per person for the whole organization.

These measures relate only to the formal act of communication, without reference to content. Innovation theory suggests, however, that openness or frankness in the expression of feelings and evaluations, a subjective freedom of communication, is very important to innovation. Such freedom makes for much fuller communication and thereby enriches inputs, and also reduces the dangers of innovation, especially the dangers to the ego or self-image of the individual. This kind of atmosphere is part of the notion of "freedom to innovate." Much work is needed on this measurement in order to make it practical. Quasi-psychiatric sessions with each individual are not practical on a large scale. As a beginning, I would suggest a questionnaire, to be filled out by all nonclerical personnel, listing a number of possible topics of conversation and asking the respondents to rate each possible topic as either a legitimate or a nonlegitimate subject of conversation while at work. Such topics as the following might be included:

1. The competence or incompetence of one's superiors,
2. The competence or incompetence of colleagues,
3. Salaries of higher officials,
4. Salaries of lower officials,
5. The quality of various projects going on in the shop,
6. The personalities of colleagues.

I think these topics need a great deal of further thought and pilot trials, but eventually it should be possible to construct a questionnaire, one reflecting widely held taboos of modern organization and interpersonal relations, that would adequately differentiate organization units on the basis of this subjective freedom. (However, great care would have to be exercised to avoid producing an instrument that reflected simply the loneliness and interpersonal superficiality of industrial man in general.) As usual, average unit scores on the questionnaire would be compared with the average organization score in order to distinguish potentially innovative areas.

JOB SATISFACTION

Like production or innovation, job satisfaction is an output (or effect) variable and it could be used, as I have used innovation, to construct its own series of measurements. As an output variable, job satisfaction varies in some obscure way with production and possibly with innovation, and these relations could probably be explored to good effect—once we learned how to measure, to specify or describe, an organization in relevant terms. We might find that an organization high on the composite innovation index was low (or medium or high) on the composite job satisfaction index; or we might find that they varied at random. There is a great deal of frustration associated with innovativeness. I do not think we can dogmatically assert that "human nature" is innovative.

The Research Program

Let me briefly describe the kinds of research associated with this approach to organizations. Five categories of research occur to me.

First, there is the kind of research that further develops our understanding of the relations between the measurements and our concern with innovation (or production, etc.). Usually this amounts to discovery of the intervening variables between the objective measurements and the output variable, innovation. We need to acquire a better understanding of the weights to be attached to the various variables, and hence to the indices. I believe that this understanding will tend to come about once we begin this kind of measurement. As we observe the relations of the measurement variables to innovation, it will soon become obvious which measures are important, which are unimportant, and which are canceled out by contrary organizational qualities. Moreover, the most useful means of stating the results of various measurements in numerical terms should also be better established. At present it would appear that all of the indices probably should be reduced to a single system of numbers, possibly a range from -1 to $+1$.

The weakest part of this research, and the weakest part of all innovation-related research, is the definition and subsequent measurement of our output variable, innovation. For the time being we should be pragmatic and use what seems best of the kinds of data that are available in the immediate inquiry—inventions, patents, publications, volume of in-and-out communications (everything has an ideational, hence communicable, form), evaluation by outside experts, internal agreement on level of innovation, evaluation by inside knowledgeables, relative speed of adoption, and perhaps a few more. Chris Argyris believes that the measurement must ultimately be psychological (e.g., level of ego-riskiness). In my opinion, however, his definition of innovativeness does not meet common sense needs for perceiving objective evidence of novelty and change, and it depends upon a rather impractical methodology and observational skill. Even so, the use of content analysis of verbatim transcripts of meetings sounds promising and may tell us much about an organization, if the effects of the chairman's personality can be neutralized.

This kind of research fills out our organizational theory of innovation. Some of it can be done by traditional experimental methods, using *ad hoc* groups of student subjects. Suppose, for example, we wished to throw further light on the question of the richness and variety of subject matter inputs as it relates to integrative versus aggregative departmentalization. We could construct a complex task that could readily be broken down into x phases or steps. In one group of subjects, each individual is required to perform a single step or phase over and over again (that is, each subject becomes a "specialist"). In a second group, each subject performs the whole task (i.e., all phases) by himself. After several task cycles have been completed by both groups, a problem concerning the task is put to the two groups, first to the individuals composing them and then to the groups as groups, and both individuals and groups are scored according to their problem-solving output (number and novelty of suggestions within a set time). The same problem is also given to a third batch of subjects, a control group whose members are wholly unfamiliar with the task. According to the theory of organizational innovation we would expect the first group to achieve the highest group score, and the members of the second group to get the highest individual scores. These two effects are maximized, theoretically, when a highly trained group of professionals (high individual scores) is brought together in an integrative group, like the first experimental group (adding high group to high individual scores).

One more example of this kind of research should suffice. The theory leads us to expect most by way of innovation from professional specialists. The belief persists, however, that this expectation is unwarranted, and that we should be training "generalists," who are—as "everyone knows"—especially good at solving problems and creating new policy (i.e., innovating). We assume that those who hold to this belief do not think of "generalists" as persons who have mastered all knowledge *or* as persons who have a superficial smattering of many fields (such as might be acquired by taking the introductory undergraduate course in dozens of fields). What, then, is the empirical content of the idea of the generalist? I think there can be only one empirical meaning to this notion. "Generalist" refers to a relationship—the relationship of *comparative* ignorance of the subject under discussion. Thus, in relation to the President, the Secretary of Defense is a military specialist, but in relation to members of the Joint Chiefs of Staff he is a generalist and they are the specialists. They, in turn, are generalists in relation to the heads of the various commands and branches, who in turn are specialists in relation to them. Etc.

The empirical question is, then: what is the nature of the contribution made to the solution of a problem in some area of activity by the person in the problem-solving group who is the most ignorant with regard to the field? I believe that there is such a contribution, one that is perhaps related to freedom from Veblen's "trained incapacity." It would be very easy for social psychologists to design an experiment, again using *ad hoc* groups of students, to test this belief and, if it proves to be correct, to discover the precise nature of the "generalist" contribution. Many fruitless debates might be resolved by such a simple piece of research.

A second kind of research is that involved in taking the measurements themselves, and developing new and better ones. It is hoped that eventually much of this can be done from personnel files, but an occasional questionnaire or interview question must still be asked (and will have to be for some time to come, I imagine).

A third kind of research activity, perhaps consisting mostly of statistical manipulation, is research to discover the interrelations (essentially the correlations) between the indices. In time, many of the measurements suggested here, or by others, will be factored out, having been found to be largely duplications of other measurements; if two variables always vary together, only one needs to be measured. (Thus, organizational measurement should grow more complex for a while, and then begin to become simpler.)

A fourth type of research will investigate relationships between the output variables (innovativeness, production, client popularity, employee satisfaction, survival or longevity, costliness, etc.), once we learn how to describe (measure) an organization in terms of them.

Finally, very interesting questions about boundary maintenance, about interface relations between different kinds of organizations, will become researchable. In the next and final section of this chapter I will deal with several findings of this type already suggested by research into the relations between research or R and D units and a production-dominated organizational environment.

Interface Adaptations

Boundary or interface relations arise between organizational entities that can be clearly distinguished by these measurements or indices. The magnitude of these adaptive activities should depend upon the magnitude of differences between the entities in question. For this kind of observation, composite indices are needed, since without them we cannot determine the magnitude of the differences. For our purposes, these differences are meaningful only in relation to our interest in innovation; consequently, the significance of the border activities is determined by this same interest. In other words, border activities of such a kind and magnitude exist because organizational differences of this kind (innovativeness) and this magnitude exist. The net difference on the composite innovation index is an innovation–noninnovation gradient, along which flow actions and reactions, attacks and counterattacks. The interface protective reactions of the innovative social system will be determined by the kinds of threats it faces. A production (or other) interest would reverse the perspective—the telescope, so to speak—and consider the protective adaptations of the unit(s) bordering the innovative one. From this new perspective, what was an attack would now be viewed as a defense.

The following illustrations of interface adaptations are taken from the literature on research units within the bureaucratic environment. I think it is reasonably correct to regard these findings as illustrations of innovative adaptations to an environment dominated by production interests. However, identifying a scientific research unit with an innovative unit poses a number of problems that have not yet been explored by research. A research unit may be an idea- or service-producing unit and be dominated by production rather than innovating values. It may be engaged in what Thomas S. Kuhn calls "puzzle-solving science" and actually not be especially innovative. However, these questions lie in virgin, unexplored territory and cannot be investigated until we learn how to give meaningful descriptions of organization units based upon measurements that are as objective as possible. Laboratories seem easy to identify without objective measurement. One has only to follow his nose. Consequently we have considerable data on labs.

It should become apparent that this research approach is strikingly different from both present management and present research, and that it is richly rewarding. I believe that budgeting and other controls should be studied from this perspective—as interface or boundary relations between different kinds of organizational units—rather than from the perspective of the logic of the controllers' intentions, as is now done by our new administrative elite, the econologicians.

Below are listed briefly a number of administrative behavioral phenomena that are best understood as innovative border adaptations to a production control-oriented organizational environment by research units.

1. Professionals have little contact with management or knowledge of "company goals."
2. Vagueness in definition of the research function promotes dispersion of discretion and greater autonomy. In such units you often find casualness about definition of mission; you find vagueness and indeterminacy.
3. Maintenance and promotion of the ideology of "pure science" helps to protect autonomy, especially in regard to the selection of problems, methods, and standards for solution.
4. One often finds internal power struggles, as between "researchers" and "developers" especially, to determine dominance over boundary relations—to determine which group will select the counts and the marquises.
5. One finds a strong tendency toward differentiation and segregation of research (in order to build up enough internal power to protect autonomy) and a concomitant tendency to reject the more production-oriented functions of development, technical service, and engineering. (Since the isolation of R and D *also* protects the status quo in the politically important part of the balance of the organization, it encounters little or no opposition.)
6. Structural integrity and autonomy are often promoted by the careful selection of clients, by having a great many clients so that no one of them acquires too much power, and by refusing to do certain kinds of work (e.g. development work) for clients. These protective adaptations to a "dangerous" environment are particularly noticeable in the case of independent labs, but they are practiced elsewhere when possible.
7. Vague project proposals help evade the controls of a project approval system of control.

8. Overoptimistic time and cost estimates commit the controllers to projects that might otherwise never get started.

9. A cleverly stated research proposal can make the proposal appear consistent with the values of the external control apparatus.

10. Avoidance of conflict by initiating only "safe" projects, avoiding anything risky or unpopular with the production control-oriented surroundings. This way lies surrender and loss of innovation.

11. Related to vague proposals is "bootlegging"—using resources budgeted for another project or purpose on problems of the researchers' own choice—fund-juggling in general.

12. Preemptive mobility—moving to other jobs in the organization, both horizontally and vertically—so as to alter the surrounding environment. Somewhat the same effect can be achieved by greater participation in meetings and other communication with other parts of the organization (sales, manufacturing, etc.). The importance for the stability of the research unit of establishing *Gemeinshaft* relations with other parts of the organization has often been documented.

13. Control of spokesmen, especially supervisors. This may take the form of insistence that a supervisor (and higher superiors) be a person having an appropriate professional or technical background. Beyond that, the "researchers" will probably recognize and reward only the supervisor's attachment to scientific or professional values, pointedly overlooking his administrative skill, personality, etc. (Note the rapid decline of the "generalist" or amateur administrator. Britain is beginning to get tired of hers. The general area administrator—e.g., a "District Officer"—is now chiefly an official of underdeveloped countries.) In its dealings with the outside world, the claims of the professional group can hardly be effectively presented by other than an appropriate professional.

14. The research unit will resist organization incentives (power,

status, money), fearing, rightly it seems, that the incentive system will attack its loose, adaptive structure, stressing conformance to organizational norms, interpersonal competition and associated secretiveness, and organizational values, identifications, and loyalties. By maintaining attachment to professional incentives, the research unit sustains the professional identifications of its members and attachment to professional values, and hence an area of pluralism within the monocratic structure of the larger organization. Without this pluralistic allegiance and independence, creativity and intellectuality tend to be smothered by the official Establishment, by coordination, and by insistence upon cooperation and a "positive" point of view—namely, an Establishment one.

15. Liaison arrangements to protect the border and hand-over transactions across it. If such arrangements exist, the research unit may attempt to gain control of them, as by preemptive placement, to convert them into buffers or Trojan horses.

16. Selective "forgetting." Some studies have found that successful researchers (meaning those whose results have been used considerably) remember some things and "forget" others. They remember from far in the past the idea sparkplugs and strong technical colleagues, but often cannot remember the names of their superiors. They easily "forget" things, like instructions, that might have the effect of reducing their autonomy or the free adaptive structure of the research organization.

17. Closely related to selective forgetting is selective inattention to regulations and requirements, such as those dealing with budgets, time, reports, and use of resources. In general, researchers are likely to be ignorant of many (perhaps most) formal arrangements, either by forgetting them or by not learning them in the first place. Associated with this forgetting and inattention is often an expressed administrative ineptness that functions to secure forgiveness for infractions. "Successful" researchers have been found to

be well aware of personal professional relations with colleagues, prospective users, and sympathetic representatives of sponsors. They are familiar with sources of technical understanding, insight, and inspiration and well acquainted with their own real goals and aims (in contrast to formal job descriptions).

18. The professsional association is beginning to be used to protect working (organizational) relationships, but not to any great extent. Professional "unions" have not been successful.

19. Getting research results put into use may in some circumstances be a protective device, serving to avoid pressure to convert the unit to a technical service function. (This question needs much more study.) Probably 70 per cent or more of research is never used. Some say that it is as hard to get research results put to practical use as it is to obtain the results in the first place. The reasons for this strange state of affairs are: (a) fear that use of the results will disrupt the political power equilibrium in the organization, (b) lack of scientific and technical knowledge sufficient to understand the research and its potentialities, (c) the short-run expediency proclivities of bureaucratic organizations (large, well-heeled organizations with slack are apparently more innovative), (d) parochial resistance to the influence of "outsiders," and (e) fear of the risks of change. Some of the methods cited above work to reduce this parochial resistance and prove the value of the research unit.

These illustrations of adaptive boundary behavior, extracted from the literature on R and D organizations, show that there is a rather obvious clash of interests and concerns between the adaptive, loose, and relatively free and undetermined research organization, on the one hand, and the rest of the monocratic, bureaucratic organization, on the other. Nevertheless, the border-maintenance adaptations and operations of research units so situated have not been systematically studied, and the more general study of the maintenance (or equilibrating) activities between different kinds of organizations has not even begun, because—the point is worth repeating—we cannot at present describe organizations as such with sufficient precision. However, the rather extensive organizational research program proposed in this chapter would, I believe, do much to remedy this methodological deficiency.

V

Society and Organizational Innovation

AN organization is a human behavioral system composed of some number of behavioral subsystems, among which the authority, status, and technical subsystems are crucial. These three, by determining the pattern of face-to-face communication within the organization, create a fourth subsystem, the group subsystem, often called the "informal organization." Certain aspects of these subsystems are the conscious products of decision-making in the organization. Most aspects, however, are determined by the culture in which the organization is embedded. Like family organization, though to a lesser extent, administrative organization is a cultural product. Consequently, a discussion of organizational innovativeness cannot be complete unless it takes into account the innovative implications of today's society and culture and the changes taking place therein.

Given the limitations of a book such as this, and also the limitations of human knowledge, the following discussion must necessarily be somewhat sketchy and speculative. Let us consider, first, some of the forces and trends that tend to support increased innovative behavior, and then turn to some of the forces and trends that might be expected to repress such behavior—that is, first the flexibilities and then the inflexibilities of the emerging society.

As industrialism has advanced, the rate of social change has greatly increased. This statement is not entirely tautological. Although it is true that invention breeds more invention and that at some point in this process we start using the term "industrialism," it is also true that there are deeper changes in the quality of life and institutions that are, in various ways, both causes and effects of this increase in the rate of invention. The changes that seem to be most closely associated with innovation can be conveniently summarized as an enormous increase in personal mobility—geographical, social, and psychological.[1]

The development of transportation and communications technology has been associated with a weakening of tribal and similar controls and the emergence of the individual.[2] Increasing mobility has been associated with the decline in the importance of area-based social organization and the growth of functionally based organization. When individuals are bound to some particular area, social organization is characterized by traditionalism, secrecy in communication, ascribed rather than achieved status, and concern with special privileges for members of the group rather than with functional policies. As we have seen, these are not qualities associated with innovation.

Mobility weakens area-based social organization and promotes the functional kind, often called associational, which is based on limited common interests and so requires only limited and temporary commitments. The individual pursues his interests through many specialized organizations rather than one all-purpose organization such as the family or tribe. His contacts multiply greatly and his horizons expand. He becomes a citizen of society rather than a member of a tribe or clan.

The increasing rate of invention reflects an enormous increase in knowledge, which, in turn, stimulates an increasing interest in education. In due course, education becomes compulsory and universal, and further development has only one direction in which to go: an advancing school-leaving age for everybody.

The qualities of a society's institutions change. Traditionalism gives way to rationalism; achievement replaces ascription as the relevant criterion on which distributions should be based. Institutions become functionally specific (rather than diffuse) and are characterized by impersonal relations, affective neutrality, and universalistic rather than particularistic norms of conduct. Because there are contrary forces at work, these social changes are of course tendencies, trends, or matters of emphasis rather than revolutionary transformations. However, the important point in the present context is the fact that many of these social changes are associated with increased innovativeness and capacity to change.

One of the more profound influences on the rate of change in our society is the changing nature of work. A few people have taken the position that "industrial slavery" is increasing, but that it is being compensated by growing leisure. Increasingly, they say, man will express his creative nature in leisure-time activities.[3] This viewpoint is quite conventional, in that it maintains the ancient notion that work is necessarily painful—that man must earn his bread by the sweat of his brow. However, it overlooks a number of important factors.

Richard L. Meier has estimated that, within another generation, on the basis of present trends, the labor force in Western countries will be distributed as follows: 10 per cent or less in agriculture, 10 to 15 per cent in manufacturing, 5 per cent in construction, and 70 to 75 per cent in services. Thus, if the industrial work place is becoming even less attractive for human beings than it has been in the past, this fact affects only a small and declining proportion of the work force. Even in industry, however, automation tends increasingly to give the drudgery to machines and to push people into jobs requiring more skill and training—in maintenance, in R and D, in professionalized "staff" work, etc.[4] Most jobs, however, will be in services, and they are not so readily programmed or routinized.

We are in the midst of an enormous expansion—an "explosion," as some have put it—of technical and scientific knowledge. This knowledge is used by organizations through the necessary agency of highly trained people. Thus, rapidly developing current technology has created an insatiable demand for highly trained people. Work and the labor force are being upgraded. Much of the labor force is becoming professionalized. The school-leaving age is advancing. Feudal arrangements that restrict mobility, such as social stratification, are rapidly giving ground or at least being drastically modified. An increasing percentage of the employees of organizations have had college training, and in an increasing number of organizations a majority or near majority of employees are professionals of one kind or another.[5]

It seems reasonable to suppose that the monocratic relations of bureaucratic organizations had their origin in an earlier and simpler era, when technology was so rudimentary that one man could master it, and when great inequalities in the conditions of life produced great inequalities in contributions and corresponding rewards within organizations.[6] Assuming that to have been the case, it seems reasonable to expect the professionalization of work to give rise to important changes in relationships and administrative practices within organizations. A professional has had a long period of pre-entry preparation for his work. Unlike the desk classes of the past, professionals do not come to the organization empty-handed to sell their undifferentiated time and effort to be used as the management thinks best. The work of the professional is not determined by the organization, and that work is usually a source of great personal satisfaction to him. Professionals develop associations to protect their work and their work standards from organizational opportunism and authority.[7]

Professionals tend to be oriented toward their professions rather than toward the particular organization in which they find themselves at the moment. They look to the profession not only for the definition of their work, but also for personal evaluations and the most important reward—professional recognition. They are more concerned about their growth within the profession than they are about their advancement within the organization. If the organization is perceived as an important channel of professional growth, they will identify with it and give it loyalty. Otherwise, their association with a particular organization is likely to be tenuous and temporary. Since they can practice their profession anywhere, they will go where the best professional opportunities are perceived. The net result of these considerations is that organizational authority and power over the individual is greatly muted where professionals have replaced the desk classes.

Professionalism, then, is an alternative to bureaucracy (or the market) as a social control. As a system of control it is pluralistic and collegiate rather than monocratic and hierarchical. The rewards it offers are professional recognition for increasing competence (professional growth) and the intrinsic satisfactions associated with professional work. When the two systems are brought together in an organization that employs a large number of professionals, basic changes in management practices become inevitable. In the first place, the resulting social system is pluralistic; there are now two sources of right, of command, of loyalty. Negotiation and compromise become necessary. With the decline in demand for managerial positions, promotion into greater power and status ceases to be such a powerful club. Consequently, the anxiety level of the organization (and conformism caused by such anxiety) declines, as more and more individuals within the organization look to their professional peers, rather than to their supervisors, for evaluation of their personal worth.

Considering these facts, it is obvious that many aspects of personnel administration must change, if it is to deal effectively with the new state of affairs. Personnel administration arose out of a desk-class age and, consequently, a desk-class bureaucracy. For the administration of professionals, much of it is obsolete. Position classification as we currently know it, for example, assumes specialized tasks, not specialized people, and accepts, therefore, the related notion of labor as a commodity. Professionalism, on the contrary, involves the specialization of people, not tasks. The task remains undefined. One might say that the essence of professionalism is that the worker assumes personal responsibility for both the definition of the problem and its solution, without supervisory oversight.

Professionalism is based on the concept of investment in human "capital," rather than that of labor as a commodity.[8] The cost of this capital stays the same even if it is not utilized fully. The only way to know the cost of under-utilization of professional talent is to require the user to pay the full cost of the human capital regardless of how he chooses to use it. If a medical doctor is used as a janitor, he should still be paid a doctor's salary. Until recently, personnel administration (position classification) has been implacably hostile to this point of view. One of the signs of the professionalization of work, and hence of organizational administration, is the beginning of the breakdown of desk-class position classification.

> The U.S. Civil Service Commission has developed, at the request and with the support of the National Science Foundation, a new system for identifying the kinds of work performed by scientists and engineers. This system will permit collecting data on the functions in which scientists and engineers are engaged ...
>
> The data to be collected will provide information not now available on the present and changing composition of the Federal work force ....[9]

With the great increase in the number of professional personnel in organizations one would expect to see less administration by top-down command, less unquestioning obedience, less restriction of communication, less parochialism and noncooperation of organization units. Organizations should more closely resemble societies of equals and human relations in them should, as a result, become more humane and dignified. All of these changes should make for more flexibility, variety, and acceptance of change. And all of them should, therefore, lead to greater innovativeness.

However, though it seems that the upgrading of work should give rise to an organizational atmosphere that will be somewhat more congenial to innovation and change than Weberian monocratic bureaucracy has generally been, professionalization also gives rise to certain problems that must be faced. Beyond the obvious ones of professional parochialism and the dogmatism of the expert, I wish to single out three for a little extra emphasis here. These are, first, social control of professionals, secondly, the extension of the threat of obsolescence to increasing numbers of people, and finally, a possible crisis in motivation.

As we have seen, professionalism is an alternative form of social control. The professional man is highly controlled by his professional peers by virtue of common values acquired in professional schools, by his constant association through work (and often socially) with other professionals, by his continuing education ("keeping up"), which is based largely on books and journals that only fellow professionals read. To the extent that professional education inculcates social values, they are likely to be conservative ones. Whether we call the result "social ignorance" or by some kinder term, the professional is to a considerable extent alienated from the rest of society.[10]

If the social changes occurring in our time were all of the kinds discussed earlier in this chapter, we could reasonably afford to be quite optimistic about the future, as likely to be a good time in which to live and work. However, there are other tendencies, such as the increasing social alienation of many professionals, that force us to temper our optimism—tendencies that appear to justify a somewhat more pessimistic outlook.

As was noted in chapter three, the successes of advancing industrial technology may themselves work against future technological advances by making further change more expensive in every way, and by reducing the proportion of production costs that is responsive to technical development. In the past labor costs have been most responsive in this regard. Now, however, automation is introducing a growing need for stability with regard to materials, products, skills, and markets.

The upgrading of work is itself introducing new inflexibilities. There have always been highly skilled workers, of course, but now the proportions of skilled to unskilled are rapidly changing. In the past, most work (both blue-collar and white-collar) consisted of specialized tasks that were simple enough to be learned from other workers on the job in a fairly short time. Workers themselves did not become specialized under these conditions and consequently acquired very little personal power. Other workers, able to do the job equally as well, could be hired off the street. The fact that management was virtually all-powerful made for great flexibility in administration. Work was almost completely under organization control and hence it could be stopped, started, or changed almost at will.

The current upgrading of work and workers means that more and more people learn "their" work before entering upon their jobs.[11] They have a sizable investment in skills; they are personally specialized; they constitute human capital. To the extent that these changes have occurred, work cannot be changed at will by management. It must be controlled by the skills likely to be available at the time needed. This condition forces the careful planning of change; it slows it up and makes it more costly. Expensive obsolescence of persons is added to the obsolescence of equipment and processes.

The rapid increase in knowledge makes more and more people liable to personal obsolescence—a fearful possibility that is no longer limited to blue-collar workers. From this fact stem two results that may be bringing us into a crisis in motivation. First is the need to keep studying in order to stay abreast of one's field. As someone has put it, we are entering an era of continuing education. For a growing number of us it is no longer possible to consider our schooling as "over" at age twenty-two (or twenty-five or thirty), in the expectation of there being nothing left to do but reap and enjoy its benefits for the rest of our lives. Nowadays we must keep going back to school, literally or figuratively, in order to "keep up." Organizational administration will have to adopt some entirely new practices, therefore, such as paying employees to attend meetings and perhaps even providing for paid sabbaticals.[12] It also seems quite likely that some people, and perhaps many, will not care to struggle in this way all of their lives, and that many of such mind will become adult "dropouts," losing all interest in both organizations and the process of "achievement" or "success."

At the same time that the phenomenal growth of knowledge in our day makes it increasingly difficult to keep up with one's specialty, the increasing codification or formalization of knowledge is reducing the significance of experience and hence the significance of age. The young man just graduated from college is likely to be better prepared, in certain critical respects, than the old grad with years of experience in a given profession, and the younger may well command the higher salary and perhaps even be given more authority than the older.[13] One can no longer count on a venerable old age, with increasing responsibility, respect, and remuneration.

These two conditions—the need to keep studying and the doubtful promise of the mature years—may well create a motivation crisis. Many a good man may ask, "Why struggle?" and will seek meaning and satisfaction in noncareer, non-organizational settings (families, vacations, hobbies, community services, etc).[14]

I expect that institutional adaptations to this problem will take the form of offering various guarantees and protections to the aging individual. Otherwise, too few individuals are likely to be willing to undergo the long period of personal sacrifice that is involved in acquiring a high level of specialized training. Unfortunately, however, such adaptations will tend to introduce inflexibilities throughout our occupational life.

Another progressing inflexibility of our mass organizational society is bureaucratic centralization. There are two general classes of causes for this increasing centralization: (1) the personal needs of persons in positions of power, and (2) technical development. The more decentralized an action arena, the more decision makers must predict the conditions of action and, especially, the behavior and responses of other relevant people. By enlarging the area of administrative control, the decision maker brings many of these predicted conditions under administrative control, and thus he reduces his own anxieties. The results of competition may be good for consumers, but competition itself is not especially congenial to human nature.

Technical development, because of the specialization involved in it and the increasing cost of its products (new skills, new processes, new equipment), leads to increased centralization, for it seems important that the new and expensive equipment and skills be kept fully occupied. The result is achieved by placing control over them at a "higher" level in the organizational hierarchy, thereby increasing the number of people dependent upon them—enlarging the market, so to speak.

Technological development, thus far, has continually increased the cost of capital equipment and has, therefore, continually increased administrative centralization. It is theoretically possible, at least in science fiction, that we may have a real scientific and technological breakthrough that could reverse this process, encouraging administrative decentralization. An inexpensive vest-pocket computer, for example, might have a decentralizing effect. However, I think it safer to expect technical development to continue to stimulate further centralization.

In a centralized system, only those with authority at the center can legitimately innovate. There are a number of forces, however, including the increasing amount of capital controlled, that tend to give rise to feelings of anxiety in those occupying major authority positions in large centralized systems. The way to reduce or at least manage such anxiety is to encourage or enforce conformism. Innovation, however, is a by-product of freedom—a true freedom in which the individual has such a sense of personal security that he is not afraid to make choices. It is by innovation, indeed, that we recognize freedom in an organization. The free person is one who is not afraid to do something different—something not dictated by authority, the group, tradition, or personality.

Herbert Simon has given us a neat phrase to summarize the most recent and most powerful thrust toward further administrative centralization: "the new science of management decision."[15] This centralizing complex is composed of a number of things, the most important of which are rapid development of data-processing technology and the access to administrative power of applied mathematicians, of "econologicians." Members of this new elite sincerely believe that we not only can, but that we should convert *all* organizational decisions into what Thompson and Tuden have called the "computational type."[16] As I pointed out in chapter three, this group shares the scientist's optimistic belief that science can solve social problems. It is, therefore, highly managerial and seeks centralized power so that mankind can finally be wisely governed.[17]

Because of strong personal needs to control within organizations —the felt needs of those "in charge"—I expect that in most organizations whatever can be controlled, will be. Automation has been converting production organizations into single physical systems. The "new science" is now in the process of converting them into single decision-making systems. Thus we hear more and more talk about the possibility of converting a whole organization—any organization—into a single system that can be integrated and directed like a tool or weapon. An advertisement in the program for the 1965 meetings of the American Society for Public Administration states:

> The Univac 418 extends space age . . . control instantly, over geographical and jurisdictional areas, at low cost. It coordinates the functions of separate departments through a common, time-shared computer system performing in instant request-and-reply cycles.

The management of men seems to be becoming very efficient indeed!

Implicit in this process is the loss, or at least the drastic reduction, of individual autonomy—of freedom and security. In the judgment of those in control, the "new science" may be said to serve some "higher purpose" (such as "survival"), but let no one doubt the cost that must be paid. Strangely enough, the econologicians, for all of their cost-benefit analysis techniques, seem disinclined to spell out these costs for us.

Freedom has not been protected by abstract philosophy so much as it has by managerial ineffectiveness (or inefficiency, if you prefer). John Milton's notion that men are creatures who *need* freedom is not one that appeals to the modern econologician. And John Stuart Mill's argument that freedom is the path to truth only works where there is at least a measure of humility concerning how much we know and how well we know it. When we begin to think we have the final answers, or at least a foolproof technique for getting them, there is not much point in allowing freedom to a bunch of untutored ignoramuses.

I suspect that man has enjoyed what freedom he has had because his masters were simply *unable* to do anything about it, either because of power equilibrium or administrative ineffectiveness. The masters' ineffectiveness in data collection, storage, retrieval, and transmission has been especially important in this respect. The life of the individual was compartmentalized, and oftentimes he could fail or stub his toe in one compartment without being hurt in another. A man usually had, and many men always had, backstage areas to which he could retreat when things got too bad. Now, however, our developing data-processing technology increasingly threatens this defense of the individual. As compartments disappear, the backstage disappears. A failure in one area of life can cause us to fail in another, and perhaps in many or all.

In the past, as Barnard has said, organizational departmentalization resulted at least partly from inefficiencies in communication.[18] Communication through such a departmentalized system requires memos, meetings, telephone conversations, etc. Information and cooperation can be withheld. Negotiation, bargaining, and compromise are essentials in this lumpy process. And within the process, the individuals can hide and can, perhaps, meet some personal needs that would not otherwise be met. Since, owing to administrative ineffectiveness, his cooperation is required, the individual is afforded opportunities to rectify, at least partially, some of the inequities that inevitably arise out of any formally designed system. Thus, the inequities of the income tax law are met by minor cheating. The hope of the administrative data processors is to eliminate the need for individual cooperation and hence the individual's ability to cheat. As they are increasingly successful in doing this, the inequities will become—or will seem to be—increasingly oppressive.

And who will stop this development? Milton and Mill will not stop it. Can one argue in favor of administrative inefficiency, of the right to refuse cooperation, of the right to cheat? The ultimate outcome of present trends, if nothing intervened, would be elimination of the "backstage" and the individual indeterminancy (freedom) it affords. A man would be front and center all of the time. Management—those who controlled, but were not controlled by, the data-processing system—would be completely effective. Freedom would all but come to an end and so would organizational innovation—except for occasional amendments in the system design.

This picture is, of course, a bit overdrawn. I do not believe that such an Orwellian tyranny will actually come into being.[19] However, failure to be alert to the possibilities and tendencies would be folly. Forewarned, we can hope to frustrate the controllers by various kinds and degrees of noncompliance, which sociologists diplomatically call "adaptations," such as those described in chapter four.

A source of strength in the American society, but also one of its weaknesses, has been its strong orientation in favor of "getting things done." This orientation, which seems so ideally suited to the needs of people in managerial or governing roles that it may aptly be termed "managerialism," may give rise to such phantasies as the widely held belief that the "problems" of "underdeveloped" countries could be "solved" by the application of "a little American managerial know-how." Most administrative studies and teaching in the United States have long been dominated by managerialism, and in my judgment this orientation is largely responsible for the fact that Public Administration, as a subdiscipline within Political Science, is so anemic. The managerial orientation has saddled it with an old-wives empiricism that has alienated many young graduate students.

Managerialism has been reinforced by the cult of success. Success has come to be defined, in America, largely in terms of moving up a managerial hierarchy. The dominant organization theory, the monocratic bureaucracy of Max Weber, has further reinforced managerialism by ascribing all organizational behavior to the decision (and command) of a manager.

Notwithstanding the claims of managerialists, however, it is not at all clear that administration has been of vital importance in the economic development of the United States. Among students of the matter, historically oriented and otherwise, there is a powerful and respectable line of argument that attributes American economic development, and especially advances in industrial productivity, almost entirely to technological change. It is suspected by some that the rapid expansion of administrative overhead in recent years has not contributed to production efficiency,[20] but, on the contrary, that increased productivity, having gone into the wages of labor and the salaries of management rather than into lower prices, has been responsible for the expanded overhead.[21]

As a result of the dominance of the managerial orientation, conscious and unconscious, recent administrative developments have been marred by a dangerous bias. For one thing, the basic motivation for these developments has been the needs of the commander—the top person in an authority role relationship. Recent "improvements" in data-processing have been seen as improvements largely from this command perspective. They help a manager *manage*, a controller *control*, a governor *govern*—and in each case more efficiently. Even the recent calls for better handling of social and political data, in a "system of national social accounting," are motivated by a desire for better (i.e. more efficient) management or governing of the society.[22]

But what of the viewpoint of the person who is subject to domination in an authority relationship—the subordinate, the citizen, the client, the patient? Relatively little has been done to improve the processing of the kinds of data that are needed for decisions in subordinate roles. The prevailing tendency is to increase and improve the means for control and regimentation of society, to strengthen the hands of those in central authority, while not doing the same for those who, in an ostensibly democratic, laissez-faire society, should ultimately be controlling them. More and more, the subordinated person in an authority relationship is an open book, while the other party to the relationship, the "authority" in charge, is shrouded in impenetrable secrecy.[23] In this age of exploding informational technology, there is urgent need for new codes of behavior and legal concepts to protect the privacy of the individual —including the one who is *not* in charge now, and who probably will not be, ever.

One of the subtler dangers of managerialism is related to the growing professionalization of work and hence of organizations. Two kinds of interpersonal relation dominate the modern economic system—the "cash nexus" relation of the market and the contractual relation of bureaucratic work. Both are very limited and both are subject to the rule of caveat emptor: "You get what you pay for—and it's your own hard luck if you get cheated." Management has developed tools designed for dealing with these relationships (e.g., the rational scanning of alternatives on a cost-benefit basis), but the fact remains that these relationships will not work very well with professionals, for two reasons.[24]

For one thing, the client or manager, although completely dependent upon the professional's contribution, is seldom if ever capable of evaluating the merits of that contribution, and he is, therefore, particularly vulnerable to possible exploitation by the professional. Hence the laws of market and contract will not work. The professional must retain a continuing interest or "property" in his contribution; it must remain to some extent a gift.

In the second place, the organization (or client) needs services from the professional that necessarily go beyond the terms of the contract—services such as innovation, discovery, invention, assumption of personal responsibility for both defining and solving the problem. With more and more professionals being forced into the large bureaucratic organization in order to practice their calling, and thus being made subject to managerialism to some extent, there exists the real possibility of a more or less gradual deterioration of professional integrity and ultimately of science.[25] This deterioration may be delayed, however, and possibly even prevented (except in isolated particular cases), if enough professionals enter the managerial positions.

It seems quite likely that a new understanding of the managerial role in personnel administration is now called for. The psychologist Herzberg provides us with a clue as to what this role should be in a professionalized future.[26] Herzberg and his colleagues, investigating the question of what people wanted from their jobs, found a group of factors that were clearly related to "good" feelings about work—the "satisfiers" (or what the investigators also term the "motivators"), and another, quite different group of factors related to "bad" feelings about work—the "dissatisfiers." The satisfiers were all related to the job itself, and the secret of their importance was that they provided the individual with a sense of personal growth and self-actualization. The dissatisfiers were all related to the job "context," and the secret of their importance was that they gave rise to a feeling of being unfairly treated. Significantly, the factors that caused dissatisfaction were the traditional tools of management: company policy and administration, technical aspects of supervision, social aspects of supervision, and working conditions. The mere absence of dissatisfiers did not give rise to any sense of satisfaction. Neither did the absence of the satisfiers cause feelings of dissatisfaction. The two groups, satisfiers and dissatisfiers, were unrelated. Injustice made the workers unhappy, but they apparently took justice for granted. Man has a sense of injustice, it seems, but not a sense of justice.

Herzberg and his colleagues interpreted their findings to mean that management should concentrate its efforts on improving and enlarging jobs so as to make them more satisfying. We have seen, however, that as professionalization advances management is rapidly losing its power to define work, and that work is increasingly being defined by pre-entry processes and controlled by professional associations. What Herzberg's findings tell us, therefore, is this: the personnel administration function of management in the future should be limited, for the most part, to avoiding, and if possible eliminating, unfairness. This conclusion may be rather startling to those who recall that Plato, more than two thousand years ago, maintained that the proper purpose of government was to secure *justice*. But perhaps "avoiding unfairness" could be interpreted as a significant step in that direction.

Will organizational life in the future be conducive to intelligence, creativity, and the self-actualization of the individual through work? The answer is far from being clear. Evidence can be found for either an optimistic or a pessimistic prediction. Perhaps this means that we have a choice. If we do, I opt for self-actualization through work. The idea that we should give up the work-place as a lost cause and seek self-actualization in leisure-time activities has a fatal flaw. Leisure time needs are administered to, for the most part, by members of bureaucratic organizations. Can slaves administer to the needs of free men? Surely not. A bureaucratically administered leisure would be the crowning horror of a thoroughly bureaucratized society.

Appendix

Industrial and Preindustrial Culture Compared In Terms of Qualities Relevant to Administration

INDUSTRIAL	PREINDUSTRIAL
1. Rational, Secular	1. *Traditional and Sacral*
2. Functionally Specific	2. *Functionally Diffuse*
3. Impersonal Relations	3. *Personal Relations*
4. Universalistic	4. *Particularistic*
5. Achievement-oriented	5. *Ascription-oriented*
6. Separation of Person and Office	6. *Personal Appropriation of Office*
7. Egalitarian	7. *Stratified, Deferential*
8. Affective Neutrality	8. *Affective Involvement*
9. Fiscal Precision and Integrity	9. *Fiscal Indeterminacy*
10. Routinization	10. *Ritualization*
11. Functional Dispersion	11. *Areal Dispersion*
12. Professional Preentry Preparation	12. *Amateurism, Apprenticeship*
13. Informality (Superficially Friendly)	13. *Formal (an Etiquette of Avoidance)*
14. Technologically Complex	14. *Technologically Simple*
15. Organizationally Complex	15. *Simple Hierarchical*
16. Clientele Involvement	16. *Clientele Alienation*
17. Many Functions	17. *Few Functions*

Industrial	Preindustrial
18. Less Scope of Authority, More Weight	18. *Less Weight of Authority, More Scope*
19. Job as Sole Source of Income	19. *Multiple Sources of Income*
20. Assured Conditions of Work (Tenure and Pensions)	20. *Nonassured Conditions of Work*
21. Policy Orientation	21. *Position and Place Orientation*
22. Guilt Culture	22. *Shame Culture*
23. Lateral Controls by Interested Citizenry	23. *Hierarchical Controls Paramount*
24. Low to Medium Prestige for Government Employment	24. *Highest Prestige for Government Employment*
25. Political Neutrality of Administration	25. *Political Involvement of Administration*
26. Elaborate Record Keeping	26. *Rudimentary Record Keeping*
27. Nation State (Popular Base)	27. *Universal Empire (Power Base)*
28. The "People" as the Owner of Administrative Resources	28. *The King as the Owner of Administrative Resources*
29. Doctrinal Basis of Political Authority	29. *Rank Basis of Political Authority*
30. Associationally Organized Citizenry	30. *Communally Organized Citizenry*
31. Extrabureaucratic Interest Articulation and Aggregation	31. *Bureaucratic Interest Articulation and Aggregation*
32. Sensitivity to Public Demands	32. *Insensitivity to Public Demands*
33. Communication: Facile, Functional, Universalistic, Egalitarian	33. *Communication: Difficult, Ceremonial, Particularistic, Deferential*
34. Great Mobility	34. *Little Mobility*
35. High Literacy	35. *Low Literacy*
36. Integrated Officialdom	36. *Officialdom a Group Apart*
37. Mobilized and Assimilated Citizenry	37. *Nonmobilized and Nonassimilated Citizenry*

APPENDIX

INDUSTRIAL	PREINDUSTRIAL
38. Strong Occupational Differentiation, Weak Rank Differentiation	38. *Strong Rank Differentiation, Weak Occupational Differentiation*
39. Minimal Personal Obligations	39. *Extended Personal Obligations*
40. Contractual Relations	40. *Status Relations*
41. Very High Per Capita Income	41. *Very Low Per Capita Income*
42. Considerable Equality in Income	42. *Great Inequality in Income*
43. Many Radios and Newspapers per 1,000	43. *Very Few Radios and Newspapers per 1,000*
44. Religion Superficial for Most People	44. *Influence of Religion Deep, Constant, Pervasive*
45. Nuclear Family	45. *Extended Family*
46. Low Birth and Death Rates	46. *High Birth and Declining Death Rates*
47. Declining Influence of Experience and, therefore, of Age	47. *Emphasis on Experience and, therefore, on Age*
48. Continuing Education	48. *Prework Education Only (if any)*
49. Innovative	49. *Static*
50. Cosmopolitan	50. *Provincial*
51. Solidarity Based on Interdependence	51. *Solidarity Based on Common Conscience*

Notes

CHAPTER I

1. *Modern Organization* (New York: Alfred A. Knopf, Inc., 1961).
2. To the same effect, see Norman Kaplan, "Some Organizational Factors Affecting Creativity," in Charles D. Orth, III, Joseph C. Bailey, and Francis Wolek, eds., *Administering Research and Development: The Behavior of Scientists and Engineers in Organization* (Homewood, Ill.: Richard D. Irwin, Inc., 1946), p. 103.

CHAPTER II

1. Graham Wallas, *The Art of Thought* (New York: Harcourt Brace, 1926), pp. 413 ff.
2. See especially Gary A. Steiner, "The Creative Organization," *Selected Papers Number Ten* (Chicago: Graduate School of Business, University of Chicago, 1962).
3. A fairly large part of the literature is summarized in Morris I. Stein and Shirley J. Heinze, *Creativity and the Individual* (Glencoe, Ill.: The Free Press, 1960). The relation between scientific productivity and freedom is documented in R. W. Gerard, *Mirror to Physiology: A Self-Survey of Physiological Science* (Washington: American Physiological Society, 1958); and D. C. Pelz, "Motivation of the Engineering and Research Specialist," American Management Association, *General Management Series,* no. 186 (1957), pp. 25-46.

4. The importance of this combination of problem challenge plus personal self-confidence is extensively documented in David C. McClelland, *The Achieving Society* (Princeton, N.J.: D. Van Nostrand Company, Inc., 1961).

5. Herbert A. Simon, *The New Science of Management Decision* (New York: Harper & Brothers Publishers, 1960), p. 39.

6. Some of this evidence is reviewed and analyzed in Peter M. Blau and W. Richard Scott, *Formal Organizations* (San Francisco: Chandler Publishing Company, 1962), ch. 5. For a more complete survey, see Irving Large, David Fox, Joel Davitz, and Marlin Brenner, "A Survey of Studies Contrasting the Quality of Group Performance and Individual Performance, 1920–1957," *Psych. Bul.*, vol. 55, no. 6 (November 1958), pp. 337–72.

7. Blau and Scott, *Formal Organizations*, ch. 5.

8. See B. Klein, "A Radical Proposal for R and D," *Fortune*, May 1958, p. 112; and B. Klein and W. Meckling, "Application of Operations Research to Development Decisions," *Operations Research*, vol. 6 (1958), pp. 352–63.

9. "Where the norms of informal groups do not favor innovation. . . . innovators will not be found to be well-integrated members of such groups."—Elihu Katz, "The Social Itinerary of Technical Change: Two Studies of the Diffusion of Innovation," in *Studies of Innovation and of Communication to the Public*, Studies in the Utilization of Behavioral Science, vol. 2 (Stanford, Calif.: Institute for Communication Research, Stanford Univ., 1962), pp. 3–35, at p. 25.

10. "Performed in a variety of settings by different researchers, these experiments [evaluating brainstorming] have most consistently demonstrated the superiority of individual over collective effort in the number of ideas generated on an issue, the quality of such ideas, or both."—Fremont J. Lyden, "Brainstorming and Group Problem-Solving: The Same Thing?," *Public Administration Review*, vol. 25, no. 4 (December 1965), p. 333. See also M. Dunnette, "Are Meetings Any Good for Solving Problems?," *Personnel Administration*, vol. 27, no. 2 (March–April 1964), p. 16.

11. Plato thought a system of social control must be founded on a "royal lie." A current discussion of the same problem arises from the study of judicial behavior. Those who work closely with the legal institutions, such as members of the bar, expect judges to be impartial —to apply neutral standards and rules to contesting partisans who

come before them. Modern behavioral research finds, however, that judges decide according to attitudes and preferences and group pressures. Yet the influence (social control) exercised by legal institutions depends upon a belief in the former concept of the impartial judge. Traditionalistic jurisprudence, therefore, can be looked upon as a necessary myth, a modern "royal lie." See Martin Shapiro, *Law and Politics in the Supreme Court* (New York: The Macmillan Company, 1964), ch. 1. He writes: "Standards, when they can be found, are weapons useful to the Court in preserving the judicial myth and urging the Justices to greater boldness." (p. 28).

12. For a fuller discussion of this stereotype, see Victor A. Thompson, *Modern Organization* (New York: Alfred A. Knopf, Inc., 1961).

13. See Richard M. Cyert and James G. March, *A Behavioral Theory of the Firm* (Englewood Cliffs, N.J.: Prentice-Hall, Inc., 1963), pp. 27-28.

14. Tom Burns and B. M. Stalker, *The Management of Innovation* (London: Tavistock Publications, 1959), See also Gerald Gordon and Selwyn Becker, "Changes in Medical Practice Bring Shifts in the Pattern of Power," *The Modern Hospital* (February 1964).

15. See Robert V. Presthus, *The Organizational Society* (New York: Alfred A. Knopf, Inc., 1962).

16. "The spiraling cost of supplemental benefits in business and industry appears to reflect misguided and futile attempts to motivate through maintenance factors. . . . Competition among companies to outdo each other in the realm of maintenance factors . . . fails to increase productivity and probably contributes to the pricing of American products and services out of the world market."—M. Scott Meyers, "The Management of Motivation to Work," an unpublished report on a motivation research project at Texas Instrument Co. See also Frederick Herzberg, Bernard Mausner, and Barbara Snyderman, *The Motivation to Work* (New York: John Wiley & Sons, Inc., 1959).

17. "Men of quite varied ability enter administration because they seek advancement. Sometimes a strongly driven scientist seeks the salary, prestige, and power that go with hierarchical rank. Such a man may or may not enjoy science, but he wants the obvious marks of success. . . . On a more modest level is the man of decent but limited talents, who wants a promotion and finds it available only by entering administration. . . . At all levels of competence, then, are men who

give up research to gain the rewards, modest or substantial, which administration offers them."—Lewis C. Mainzer, "The Scientist As Public Administrator," *The Western Political Quarterly*, vol. 16, no. 4 (December 1963), pp. 814–29, at p. 816. This study was based on interviews with a large number of federal scientists-turned-administrators. Of the federal executives at grades GS-14 and above, only about one in forty-five has had any college training in public administration. Derived from table 42B, p. 361, of W. Lloyd Warner, Paul P. Van Riper, Norman H. Martin, and Orvis F. Collins, *The American Federal Executive* (New Haven, Conn.: Yale University Press, 1963).

18. See Rollo May, *The Meaning of Anxiety* (New York: The Ronald Press Co., 1950), especially pp. 181–89; and A. H. Maslow, *Motivation and Personality* (New York: Harper & Brothers, 1954), ch. 5.

19. Nigel Walker, *Morale in the Civil Service: A Study of the Desk Worker* (Edinburgh: The University Press, 1960).

20. Karl Mannheim, *Ideology and Utopia* (New York: Harcourt, Brace, & Co., 1936), pp. 105–106. See also Victor A. Thompson, *The Regulatory Process in OPA Rationing* (New York: King's Crown Press, 1950), ch. 7.

21. Walker, *Morale in the Civil Service*.

22. See Everett C. Hughes, *Men and Their Work* (Glencoe, Ill.: The Free Press, 1958); Theodore Caplow and Reece J. McGee, *The Academic Marketplace* (New York: Basic Books, 1958), p. 85; Leonard Reissman, "A Study of Role Conceptions in Bureaucracy," *Social Forces*, vol. 27 (1949), p. 308; Alvin W. Gouldner, "Cosmopolitans and Locals," *Administrative Science Quarterly*, vol. 2 (1957–1958), pp. 281–306, 444–80; and Harold L. Wilensky, *Intellectuals in Labor Unions* (Glencoe, Ill.: The Free Press, 1956), pp. 129–44. These studies are primarily concerned with the professional orientation. The bureaucratic orientation is implicit as a contrast.

23. Warner, Van Riper, Martin, and Collins, *American Federal Executive*, p. 155; W. Lloyd Warner and James Abegglen, *Occupational Mobility in American Business and Industry* (Minneapolis: University of Minnesota Press, 1955).

24. Burns and Stalker, *Management of Innovation;* Melville Dalton, "Conflicts Between Line and Staff Managerial Officers," *American Sociological Review*, vol. 15 (1950), pp. 342–51.

25. See R. M. Cyert, W. R. Dill, and J. G. March, "The Role of Expectations in Business Decision-Making," *Administrative Science Quarterly,* vol. 3, no. 3 (December 1958); and Cyert and March, *A Behavioral Theory,* ch. 4.

26. Burns and Stalker, *Management of Innovation.* "Go down the list of commercial inventions over the last thirty years: with very few exceptions the advances did not come from a corporation laboratory."—William H. Whyte, Jr., *The Organization Man* (Garden City, N.Y.: Doubleday Anchor Books, Doubleday & Company, Inc., 1957), pp. 237-38. Whyte lists, as examples, kodachrome and the jet engine. It is common knowledge that most of the scientific and technological development of the past twenty-five years has grown out of military needs. Bureaucratic organizations may gradually perfect existing inventions, but "Revolutionary developments still come from outsiders." —Morris Janowitz, *The Professional Soldier* (Glencoe, Ill.: The Free Press, 1960), p. 28. His examples include the jet engine, the I.C.B.M., and change in air raid warning and antiaircraft defense systems. See also Walter Millis, *Arms and Men: A Study in American Military History* (New York: G. P. Putnam's Sons, 1956), p. 350.

The arguments for dispersing science throughout the federal administration rather than concentrating all science-supporting activities in a single national agency like the National Science Foundation are also the arguments for dispersing R and D rather than segregating it in a single unit of an organization. ". . . we need to see that it [science] is infused into the program of every department and every bureau," as Don Price said. To which we add: "And every division and every branch," etc. See Don Price, *Government and Science* (New York: New York University Press, 1954), p. 63, and ch. II in general. The Chicago Board of Education recently established the position of "Assistant Superintendent for Integration." But, of course, integrationists want integrative thinking throughout the school board organization— not just in a Department of Integration—even as Mr. Price urged scientific thinking throughout the federal government—not just in a "department of science." This "pinpointing of responsibility" is a typical amateur reaction to organization problems. It is not based on hard knowledge of organizational phenomena.

27. James G. March and Herbert A. Simon, *Organizations* (New York: John Wiley & Sons, Inc., 1958), pp. 150-54; Victor A. Thomp-

son, *Modern Organization*, part 2; Eliot D. Chapple and Leonard R. Sayles, *The Measure of Management* (New York: The Macmillan Co. 1961), pp. 18-40.

28. Burns and Stalker, *Management of Innovation*; Melville Dalton, "Conflicts."

29. Burns and Stalker, *Management of Innovation*.

30. Klein and Meckling ("Application of Operations") reach this conclusion with regard to new weapons development. The important point about their argument is that it applies to production interests as well as an interest in innovation. As Hirshman and Lindblom summarize it: "They allege that development is both less costly and more speedy when marked by duplication, 'confusion' and lack of communication among people working along parallel lines." Albert O. Hirshman and Charles E. Lindblom, "Economic Development, Research and Development, Policy Making: Some Converging Views," *Behavioral Science*, vol. 7, no. 2, (April 1962), pp. 211–22. See also Albert O. Hirshman, *The Strategy of Economic Development* (New Haven: Yale University Press, 1958); and David Braybrooke and Charles E. Lindblom, *A Strategy of Decision* (New York: The Macmillan Company, 1963).

31. Apparently only about 25 percent of the suggestions turned in have any usefulness at all. Of the suggestion systems established, "Most die in infancy, a few become moribund in youth, some survive. . . ."—Norman J. Powell, *Personnel Administration in Government* (Englewood Cliffs, N.J.: Prentice-Hall, Inc., 1956), pp. 438–44. Powell believes that suggestion box systems are better than no communication with the rank and file at all. Because of disputed authorship of suggestions, the TVA decided to give only group (non-cash) awards.

32. See Thompson, *Modern Organization*, pp. 129–37.

33. See Kurt W. Back, "Decisions Under Uncertainty," *The American Behavioral Scientist*, vol. 4, no. 6 (February 1961).

34. Some of this evidence is reviewed in Blau and Scott, *Formal Organizations*, ch. 5.

35. The evidence is reviewed in Cecil A. Gibb, "Leadership," in Gardner Lindsey, ed., *Handbook of Social Psychology* (Reading, Mass.: Addison-Wesley Publishing Company, Inc., 1954), vol. 2, pp. 877–917. See also the leadership studies included in Dorwin Cartwright and Alvin Zander, eds., *Group Dynamics*, 2nd ed. (Evanston, Ill.: Row, Peterson & Company, 1962), part 5.

36. Andre L. Delbecq, "Leadership in Business Decision Conferences" (unpublished Ph.D dissertation, Indiana University, 1963).

CHAPTER III

1. Quoted by Rensis Likert, "Motivation and Increased Productivity," *Management Record*, vol. 18, no. 4 (April 1956), p. 128.
2. Richard M. Cyert and James G. March, *A Behavioral Theory of the Firm* (Englewood Cliffs, N.J.: Prentice-Hall, Inc., 1963), ch. 3. "It is most frequently assumed in economic analysis that the firm is trying to maximize its total profits."—William J. Baumol, *Economic Theory and Operations Analysis* (Englewood Cliffs, N.J.: Prentice-Hall, Inc., 1961), p. 193.
3. James G. March and Herbert A. Simon, *Organizations* (New York: John Wiley & Sons, Inc., 1958), p. 158.
4. "The cost of using general-purpose programs to solve problems is usually high. It is advantageous to reserve these programs for situations that are truly novel, where no alternative programs are available. If any particular class of situations recurs often enough, a special-purpose program can be developed which gives better solutions and gives them more cheaply than the general problem-solving apparatus." —Herbert A. Simon, *The New Science of Management Decision* (New York: Harper & Brothers Publishers, 1960), p. 7.
5. According to William H. Whyte, Jr., this fact helps account for the fact "that industry has a disproportionately small share of top men [in science]." See his *The Organization Man* (Garden City, N.Y.: Doubleday Anchor Books, Doubleday & Company, Inc., 1957), pp. 229 ff., at p. 229.
6. "The automation of important parts of business data-processing and decision-making activity, and the trend toward a much higher degree of structuring and programming of even the nonautomated part will radically alter the balance of advantage between centralization and decentralization. The main issue is not the economics of scale . . . . Rather, the main issue is how we shall take advantage of the greater analytic capacity, the larger ability to take into account the interrelations of things . . . ." —Simon, *New Science*, p. 45. See also Gerald Gordon and Selwyn Becker, "Changes in Medical Practice Bring Shifts in the Patterns of Power," *The Modern Hospital* (February 1964).

7. See Joseph J. Spengler, "Bureaucracy and Economic Development," in Joseph LaPalombara, *Bureaucracy and Political Development* (Princeton, N.J.: Princeton University Press, 1963), pp. 216–18. For some strange reason, Spengler believes that overrequirement is restricted to "public" bureaucracies. A vice president of the National Association of Manufacturers, Charles A. Kothe of Tulsa, Okla., disagrees with him. Speaking in behalf of the equal employment provisions of the Civil Rights Act of 1964, he said that they would force employers to reexamine "our outmoded personnel policies." He said that many of the skills required of employees had "nothing to do with their ability to do the job assigned to them." (Reported in the Syracuse *Post-Standard* (Oct. 3, 1964), p. 4.) Hundreds of intersections in Chicago are not staffed by crossing guards because so few can pass the civil service test. Of 126 women who took the exam on July 18, 1967, only 21 passed. (Reported in the Chicago *Sun-Times* [Sept. 23, 1967], p. 14.)

8. David Braybrooke and Charles E. Lindblom, *A Strategy of Decision* (New York: The Macmillan Company, 1963).

9. March and Simon, *Organizations,* ch. 6.

10. "We must try to arrange things so that the human plays as small a role as possible in the gathering of data. If the human observer has a fairly complicated role we can try to break this task up into a number of simple operations."—Irwin D. J. Bross, *Design for Decision* (New York: The Macmillan Company, 1953), p. 151. See Victor A. Thompson, *Modern Organization* (New York: Alfred A. Knopf, Inc., 1961).

11. March and Simon, *Organizations,* p. 141.

12. As I noted earlier, the *process* of creating is often more important to the creative person than the *result*, but the decisional rule governing search activities does not take this fact into account. No rule handles the problem of the consequences of the consequences of the consequences (as Braybrooke and Lindblom point out in *A Strategy of Decision*). Apparently it is assumed that the goal set values of all of these generations of consequences remain constant. Otherwise, an attempt would have to be made to predict the future indefinitely. Equally puzzling is the fact that the goal set changes as information about the consequences comes in.

13. Zvi Griliches, in his comments on Mueller's paper in National Bureau of Economic Research (NBER), *The Rate and Direction of*

Inventive Activities (Princeton, N.J.: Princeton University Press, 1962), p. 347. See also, in the same work, William Fellner, "Does the Market Direct the Relative Factor-Saving Effects of Technological Progress?," pp. 171–88, and John L. Enos, "Invention and Innovation in the Petroleum Refining Industry," p. 319.

14. S. C. Gilfillan, *The Sociology of Invention* (Federalsburg, Md.: Stowell, 1935), pp. 43–46. Gilfillan studied "19 most useful inventions introduced in the 25 years before 1913, selected by vote of *Scientific American* readers. The average intervals were: between when the invention was first thought of and the first working machine or patent, 176 years; thence to the first practical use, 24 years; to commercial success 14 years, to important use 12 years . . . ."—*The Sociology of Invention*, p. 96. On another group of "most important inventions of the generation before 1930," he found an average of 33 years from the "conception date" (patent or model date) to the date of commercial success (pp. 96–97). These averages were for the successful inventions only. Only 10 to 15 per cent of patents have paid off, he says (pp. 109–19). See also Enos, "Invention and Innovation"; Simon Marcson, *The Scientist in American Industry* (Princeton, N.J.: Industrial Relations School, 1960), pp. 109–10; Richard R. Nelson, "The Economics of Invention: A Survey of the Literature," *Journal of Business*, vol. 23, no. 2 (April 1959), p. 114; C. Wilson Randle, "Problems of Research and Development Management," *Harvard Business Review*, vol. 37 (1959), p. 133; William Kornhauser, *Scientists in Industry: Conflicts and Accommodation* (Berkeley: University of California Press, 1962), p. 68.

15. See National Science Foundation reports on R and D expenditure by industry—for example, various issues of the pamphlet, *Reviews of Data on Research and Development*, or the study, *Research and Development in Industry, 1961*. Note the very small percentage of R and D funds used for basic research: 4 per cent. Even where there are research labs there is little invention. "These organizations continue to rely heavily upon other sources of original thinking."—J. Jewkes, *et al., The Sources of Invention* (London: The Macmillan Company, 1958), p. 185. Few Nobel Prize winners have come from industrial labs since 1900.—Jewkes, *et al., The Sources of Invention*, p. 185. Except for alloys, nearly all inventions in the aluminum industry have come from outside the industry. —Merton J. Peck, "Inventions in the Postwar American Aluminum Industry," in NBER, *Rate and Direc-*

tion, p. 285. "Except for nylon, orlon, and neoprene, Du Pont's major product innovations have been based upon technology acquired from others."—Willard F. Mueller, "The Origins of the Basic Inventions Underlying DuPont's Major Product and Process Innovations, 1920 to 1950," in NBER, *Rate and Direction* pp. 323–46. S. C. Gilfillian says the revolutionary inventions in the history of the ships have been by outsiders. —In NBER, *Rate and Direction*, p. 89. Waldemar Kaempffort claims to have documented the principle that insiders only perfect —revolutionary change comes from outside. See his "Systematic Invention," *Forum*, vol. 70, no. 4 (October 1923), pp. 2,010–18, and "Invention by Wholesale," *Forum*, vol. 70, no. 5 (November 1923), pp. 2,116–22. See also Reginald A. Fessenden, *The Deluged Civilization of the Caucasus Isthmus* (Boston: T. J. Russell Printers, 1923). M. P. O'Brien, Dean Emeritus, College of Engineering, University of California at Berkeley, says: "But I'm convinced that no operating organization, through its main chain of command, will ever come up with radically new products."—"Technological Planning and Misplanning," in James R. Bright, ed., *Technological Planning on the Corporate Level* (Cambridge: Harvard University Graduate School of Business, 1962), p. 96.

16. Thompson, *Modern Organization*; Gary A. Steiner, "The Creative Organization," *Selected Papers Number Ten* (Graduate School of Business, University of Chicago, 1962); and Morris I. Stein and Shirley J. Heinze, *Creativity and the Individual* (Glencoe, Ill.: The Free Press, 1960).

17. For a description of the model of rational choice, see March and Simon, *Organizations*, esp. ch. 6. For the nonrational processes involved in invention see many of the works summarized in Stein and Heinze, *Creativity and the Individual*. On the nonrational model, see Kurt W. Back, "Decisions Under Uncertainty," *The American Behavioral Scientist,* vol. 4, no. 6 (February 1961), pp. 14–19.

18. James R. Bright, *Automation and Management* (Boston: Graduate School of Business Administration, Harvard University, 1958), p. 87. The coalition (pluralistic) nature of organizations prevents getting all the facts. To attempt to do so would immobilize the organization with internal conflict. See Cyert and March, *A Behavioral Theory*, pp. 77–82.

19. Bright, *Automation and Management*, pp. 85–86.

20. Ibid., p. 86. The accounting people were particularly doubtful about Ford's new 1948 engine plant in Cleveland, the most modern of its time. Ford and the other automobile manufacturers have since gone beyond.—Ibid., p. 60.

21. See the report of a conference on the subject by James R. Bright, ed., *Technological Planning*. The Industrial Research Institute holds conferences on this problem from time to time, publishing its results in the journal, *Research Management*. James W. Hackett says that one reason why IRI is becoming interested is that the National Association of Accountants "had set up a project to develop a technique for evaluating research," and hence it was important to keep them from going too far off the beam.—*Technological Planning*, p. 241.

22. Nelson, "Economics of Invention," p. 122. See also John W. Haefele, *Creativity and Innovation* (New York: Reinhold Publishing Corporation, 1962), p. 186; James B. Quinn, "Top Management Guides for Research Planning," in Bright, ed., *Technological Planning*, pp. 196–97; Jewkes et al., *The Sources of Invention, passim;* Albert H. Rubenstein, in Burton V. Dean, ed., *Operations Research in Research and Development* (New York: John Wiley & Sons, Inc., 1963), p. 198.

23. A. W. Marshall and W. H. Meckling, "Predictability of the Cost, Time, and Success of Development," in NBER, *Rate and Direction,* pp. 461–75; Burton H. Klein, "The Decision Making Problem in Development," in NBER, *Rate and Direction,* p. 492.

24. For expositions of some of the control devices, see David M. Stires and Maurice M. Murphy, *PERT (Program Evaluation Review Technique) CPM (Critical Path Method)* (Boston: Materials Management Institute, 1962); Raymond Villers, "The Scheduling of Engineering Research," *Journal of Industrial Engineering* (November–December 1959); *Research and Development Progress Reporting* (Washington: Policy and Procedures Division, Air Force System Command, USAF, 1961); John S. Harris, "New Product Profile Chart," reproduced in James R. Bright, *Research, Development, and Technological Innovation* (Homewood, Ill.: Richard D. Irwin, Inc., 1964), pp. 404–13. Donald G. Malcolm says that PERT will "provide management with a set of tools that will aid in enforcing the development plan; also a quantitative number predicting the reliability of

the item under design will be available at all stages in the development project."—"Integrated Research and Development Management Systems," in Dean, ed., *Operations Research,* p. 142.

25. The fact that a situation is not controlled by economic rationality does not mean that it is uncontrolled. For example, cost-plus contracts neutralize economic motives and convert the business-government relation into one of professional and client—the scientists and engineers of business on one side, and the public servants on the other. Presumably, the former act according to their best professional judgment of the clients' interests. The relation is professionally controlled. See Kenneth J. Arrow, "Economic Welfare and the Allocation of Resources for Invention," in NBER, *Rate and Direction,* p. 624.

26. National Science Foundation, *Reviews of Data on Research and Development,* no. 40 (September 1963).

27. Bruce D. Henderson, "Implications of Technology for Management," in Bright, ed., *Technological Planning on the Corporate Level,* pp. 245–46.

28. Gilfillan, *The Sociology of Invention,* p. 52.

29. NSF, *Reviews of Data on Research and Development,* no. 41 (September 1963).

30. Ibid., no. 40.

31. Ibid. Firms which operate in a swiftly changing technical field are forced to allocate more money to R and D as a hedge against the future. Firms with a great variety of products have a better chance of recovering benefits from the unpredictable output of research. See Richard R. Nelson, "The Simple Economics of Basic Scientific Research," *Journal of Political Economy,* vol. 67, no. 3 (June 1959), p. 302. W. Allen Wallis says that machinery manufacturers do most of the innovating in the field of automation because they serve the entire field and thus have a better chance of reaping the benefits of research. See his "Some Economic Considerations," John T. Dunlop, ed. *Automation and Technological Change* (Englewood Cliffs, N. J.: Prentice-Hall, Inc., 1962), pp. 103–13. See also Bright, *Automation and Management.*

32. NSF, *Reviews of Data on Research and Development,* no. 41.

33. Ibid., no. 40.

34. NSF, *Scientific Manpower Bulletin,* no. 18 (November 1962). New York City and Newark are average.

35. Ibid.

36. Ibid., no. 20 (March 1964); Francis Bello, "Industrial Laboratory," *Fortune*, November 1958, p. 214.

37. See Kenneth J. Arrow, "Economic Welfare . . ."; O'Brien, "Technological Planning and Misplanning"; Gilfillan, *The Sociology of Invention*, p. 101; Floyd L. Vaughan, *Economics of Our Patent System* (New York: The Macmillan Company, 1925); Fessenden, *The Deluged Civilization;* C. Roy Watson, "A Foundation Proposed," *Patent Office Society Journal*, vol. 12, no. 9 (September 1930), pp. 387–90; Samuel E. Hill and Frederick Harbison, *Manpower and Innovation in American Industry* (Princeton, N. J.: Industrial Relations Section, Department of Economics and Sociology, Princeton University, 1959), p. 60; Kornhauser, *Scientists in Industry*, pp. 22–23.

38. See Joseph Harrington, Jr., "A Look into Tomorrow," in *New Views on Automation*, Joint Economic Committee, 86th Congress,, 2nd Session, p. 44. Bright (*Automation and Management*, p. 15) says: "The average manufacturing system of 1956 can be regarded as no more than a crude assemblage of unintegrated bits of mechanism." See also W. Allen Wallis, "Some Economic Considerations," p. 110.

39. Bright, *Automation and Management, passim,* especially pp. 136–42, 225–26, 231.

40. See Back, "Decisions Under Uncertainty." The literature on individual creativity shows that it, too, must be explained by reference to a nonrational explanatory system. See Stein and Heinze, *Creativity and the Individual*.

41. See, for example, Quinn, "Top Management Guides," pp. 174 and 195; and Jewkes, *et al., The Sources of Invention*, p. 140.

42. Stein and Heinze, *Creativity and the Individual;* Thompson, *Modern Organization;* Back, "Decisions Under Uncertainty."

43. See Quinn, "Top Management Guides," p. 173. O'Brien ("Technological Planning and Misplanning," p. 97) says: "I could name some men in pretty big research laboratories whose directors have told me, 'we spend three-quarters of our time putting out fires.'" Kornhauser, *Scientists in Industry;* Tom Burns and G. M. Stalker, *The Management of Innovation* (London: Tavistock Publications, 1959).

44. Ward Edwards, "The Prediction of Decisions Among Bets," *Journal of Experimental Psychology*, vol. 51 (1955); Richard Cyert, James G. March, and William A. Starbuck, "Two Experiments on

Bias and Conflict in Organizational Estimation," *Management Science*, vol. 7, no. 3 (April 1961).

45. Marshall and Meckling, "Predictability of Cost"; Klein, "The Decision Making Problem"; Braybrooke and Lindblom, *A Strategy of Decision, passim.*

46. Jewkes, et al., *The Sources of Invention*, p. 151; Organization for European Economic Cooperation, *The Organization of Applied Research in Europe*, vol. 1, p. 29.

47. Jewkes, et al., *The Sources of Invention*, p. 151; Barker S. Sanders, "Some Difficulties in Measuring Inventive Activities," in NBER, *Rate and Direction*, p. 59; Herbert Shepard, "The Dual Hierarchy in Research," in Charles D. Orth, III, Joseph C. Bailey, and Francis Wolek, eds., *Administering Research and Development: The Behavior of Scientists and Engineers in Organization* (Homewood, Ill.: Richard D. Irwin, Inc., 1964); Herbert Shepard, "Nine Dilemmas of Industrial Research," *Administrative Science Quarterly*, vol. 1, no. 3 (December 1956), pp. 295–309; Marcson, *The Scientist in American Industry*, pp. 38–39, 97; Charles D. Orth, "The Optimum Climate for Industrial Research," in Orth, Bailey, and Wolek, *Administering Research and Development*, p. 368. The production organization, dealing only in extrinsic rewards, must promise the professional, and especially the research scientist or engineer, things it cannot supply. This produces an undertone of deceit and manipulation in research administration. Marcson says that research administration is conceived by management "as largely astuteness . . . the ability to outsmart a subordinate."—*The Scientist in American Industry*, p. 132. Even the much vaunted "free time" of some industrial scientists is not so free; ". . . the rewards go to those scientists who 'come up with' usable, practical ideas" on their "free" time. See Norman Kaplan, "Some Organizational Factors Affecting Creativity," in Orth, Bailey, and Wolek, *Administering Research and Development*, p. 109.

48. Orth, "The Optimum Climate," p. 370.

49. Marcson, *The Scientist in American Industry*, p. 100.

50. NSF, *American Scientific Manpower 1962*, p. 18.

51. Kornhauser, *Scientists in Industry*, pp. 143 ff.; Haefele, *Creativity and Innovation*, p. 181. There are many problems in introducing ladders or "success" systems in monocratic organizations, and

some say that the dual or technical ladder is declining. See R. M. Hower and C. D. Orth, *Managers and Scientists: Some Human Problems in Industrial Research Organizations* (Boston: Harvard University Press, 1963), p. 316, and references there cited.

52. As a percentage of entering university freshmen, engineers went from 10.8 per cent in 1957 to 7.3 per cent in 1961. In 1950 we graduated 52,700 engineers; in 1960, 37,000; in 1963, 33,458. We need 81,000 per year to meet domestic and foreign aid needs. In 1951, we graduated 19,600 students in the physical sciences; in 1960, only 17,100 despite a large increase in the college population; in 1963, still fewer—16,276. In 1951, we graduated 22,500 in the biological sciences; in 1960, only 16,700. The supply is dropping, but the needs are growing. NSF says we may need 85 per cent more than we are getting by 1970. Even in a field like auto mechanics we are slipping. There was 1 mechanic to each 73 cars in 1950. In 1961, there was 1 to each 87 cars. See Aaron Levenstein, *Why People Work: Changing Incentives in a Troubled World* (New York: Crowell-Collier Press, 1962), pp. 188–97. The number of physicians per unit of population has declined. With perhaps less than half of the scientists and engineers doing the work for which they were trained, it is a little hard to accept the concept of "shortages." The problem is a motivational one. With "success" defined as moving up the managerial hierarchy, we cannot expect all of the scientists and engineers to remain at their stations and accept their lot as relative "failures." See Herbert Shepard, "Superiors and Subordinates in Research," *Journal of Business,* vol. 29, no. 4 (October 1956), pp. 264–65. He found in two surveys of industrial research engineers and scientists that "over 90 per cent said that they aspired to management positions. Only about 10 per cent felt that professional achievement in science or engineering constitutes 'success.'" In a survey of over 1,300 engineers in some 200 companies, 78 per cent said they planned to shift to other fields. See *How to Attract and Hold Engineering Talent,* Executive Research Survey No. 3 (Washington, D. C.: Professional Engineers Conference Board for Industry, 1954), p. 37. Most university presidents have come up by the management route (73 per cent). Less than half have ever been professors. Between the 1899–1900 edition and the 1960–61 edition of *Who's Who,* management people went from approximately 4 in 30 to 13 in 30. Professionals in industry are generally frustrated and dissatisfied. See

David G. Moore and Richard Renck, "The Professional Employee in Industry," in Orth, Bailey and Wolek, *Administering Research and Development*, p. 55.

53. See Victor A. Thompson, "Bureaucracy and Innovation," *Administrative Science Quarterly*, vol. 10, no. 1 (June 1965), pp. 1-20; Cyert and March, *A Behavioral Theory*, pp. 36-38; and E. Mansfield, "Technical Change and the Rate of Imitation," *Econometrica*, vol. 29 (October 1961), pp. 741-66.

54. See Bross, *Design for Decision*, pp. 105-106, p. 116.

55. Everett M. Rogers, "Characteristics of Agricultural Innovators and Other Adopter Categories," in *Studies of Innovation and of Communication to the Public* (Stanford: Institute of Communications Research, Stanford University, 1962); E. A. Wilkening, "The Communication of Ideas on Innovation in Agriculture," in *Studies of Innovation*; Elihu Katz, "The Social Itinerary of Technical Change: Two Studies of the Diffusion of Innovation," in *Studies of Innovation*.

56. Mathew B. Miles, ed., *Innovation in Education* (New York: Columbia University Teachers College, 1964); E. M. Rogers, *Diffusion of Innovation* (New York: The Macmillan Company, 1962).

57. See note 53, above.

58. Raymond Villers, *Research and Development: Planning and Control* (New York: Financial Executives Research Foundation, 1964), p. 25. See also Kornhauser, *Scientists in Industry*, p. 60.

59. See Rensis Likert and Stanley E. Seashore, "Making Cost Control Work," *Harvard Business Review*, vol. 41, no. 6 (November-December 1963), pp. 96-108.

60. Villers, *Research and Development*, pp. 25-26.

61. Joseph A. Schumpeter, *Capitalism, Socialism and Democracy* (New York: Harper & Brothers, 1950), and *The Theory of Economic Development* (Cambridge: Harvard University Press, 1951). See also J. K. Galbraith, *American Capitalism* (Boston: Houghton Mifflin Company, 1952); and Gilfillan, *The Sociology of Invention*, p. 55.

62. Bright, *Automation and Management*, pp. 36-37.

63. Kornhauser, *Scientists in Industry*, p. 182.

64. William James, *Psychology: Briefer Course* (New York: Holt, 1923), p. 187; Daniel Lerner, "Toward a Communication Theory of Modernization," in Lucien Pye, ed., *Communication and Political Development* (Princeton, N.J.: Princeton University Press,

1963), p. 333, who says: "A person with high achievement may still be dissatisfied if his aspirations far exceed his accomplishments." The use of "satisficing" norms instead of maximizing ones is an adaptation to limitations imposed by reality. See Herbert A. Simon, "A Behavioral Model of Rational Choice," *Quarterly Journal of Economics*, vol. 69, no. 1 (February 1955), pp. 99–118. However, the important question is whether an organization *tries* to maximize and how hard it tries. Organizations with slack do not have to try.

65. *The Sources of Invention*, pp. 247–49.

66. Klein, "The Decision Making Problem"; Braybrooke and Lindblom, *A Strategy of Decision;* Albert O. Hirshman, *The Strategy of Economic Development* (New Haven, Conn.: Yale University Press, 1958).

67. In his comments on Klein's paper, in NBER, *Rate and Direction*, p. 498.

68. Ibid., p. 501.

69. Ibid., p. 502.

70. See Kenneth J. Arrow, "Economic Welfare . . ."; Braybrooke and Lindblom, *A Strategy of Decision;* Hugh M. Pease, "Raising Productivity in Problem Solving Operations," *Personnel Administration,* vol. 27, no. 1 (January-February 1964), pp. 39–42; and Alain Enthoven and Harry Rowen, "Defense Planning and Organization," in NBER, *Public Finances, Needs, Sources, and Utilizations* (Princeton, N.J.: Princeton University Press, 1961), pp. 369–71. Klein says: "In almost all the outstanding programs we have examined, the developer started out with a very loose definition of the system he was trying to develop and has exhibited a considerable willingness . . . to pursue multiple approaches to difficult technical problems."—"The Decision Making Problem," p. 493.

71. Klein, "The Decision Making Problem." See also Cyert and March, *A Behavioral Theory,*

72. Peck, "Inventions in Postwar American Aluminum Industry," p. 294.

73. Aaron Wildavsky, *The Politics of the Budgetary Process* (Boston: Little, Brown & Company, 1964), pp. 160 ff.

74. The situation is even worse in local government. See Municipal Manpower Commission, *Governmental Manpower for Tomorrow's Cities* (New York: McGraw-Hill Book Company, Inc., 1962).

See also the review of this book by Donald C. Stone, "Manning Tomorrow's Cities," *Public Administration Review,* vol. 23, no. 2 (June 1963), pp. 99–104.

President Johnson recently appointed members to a Council of the Administrative Conference to improve federal administration. Nearly all of the eleven members are attorneys. None is a social scientist. See the Washington *Post* (Feb. 8, 1968), p. 3.

75. Reported in the Syracuse *Post-Standard* (Apr. 10, 1965), p. 4. Donald Zylstra, in the same newspaper (Jan. 23, 1966), asserted that "Most persons who pointedly disagreed with McNamara have now left the department." McNamara was judged, apparently, by a set of conventionalized administrative criteria that had nothing to do with the substantive output of the Department of Defense. One is reminded of an aphorism that went around Washington in a past day when people in that city still had a sense of humor: "What is not worth doing at all is not worth doing well."

76. Reported in the Syracuse *Post-Standard* (Nov. 15, 1965), p. 5.

77. Thomas S. Kuhn, *The Structure of Scientific Revolutions* (Chicago: University of Chicago Press, 1962). My discussion of the physical sciences follows this brilliant book.

78. *The Two Cultures and a Second Look* (New York: Mentor Books, 1964); Kuhn, *The Structure of Scientific Revolutions.*

79. *Industrial Leadership* (New Haven: Yale University Press, 1916).

80. *Science and Government* (New York: Mentor Books, 1962), p. 32. We call it "operations research."

81. F. J. Roethlisberger and William J. Dickson, *Management and the Worker* (Cambridge, Mass.: Harvard University Press, 1939).

82. U.S. Bureau of the Budget, *Measuring Productivity of Federal Government Organizations* (Washington: G.P.O., 1964), p. 22.

83. *The Wall Street Journal* (Aug. 26, 1965).

84. *The New Utopians: A Study of System Design and Social Change* (Englewood Cliffs, N.J.: Prentice-Hall, Inc., 1965).

85. For a brief discussion of PPB see George Shipman, "Developments in Public Administration," *P.A.R.,* vol. 26, no. 1 (March 1966), pp. 77–79. See also David Novick, ed., *Program Budgeting* (Washington: G.P.O., 1965). The December 1966 issue of *Public Administration Review* is largely devoted to a symposium on the subject.

86. The organizational *function* of a manager is to act as repre-

sentative and spokesman for the organizational claim. Otherwise, under conditions of increasing professionalization, with its nonorganizational orientation, the survival of specific organizations would be inexplicable. Such an activity has to be organizationally rewarded, and it frequently involves the manager in sabotage of the organization's technical activities, as Veblen said. (*Vide* the automobile industry and the safety issue.) In most cases of a highly professionalized organization (e.g., a hospital), the organizational claim must be based on the inflexibility of its capital equipment and the value of the informal organization.

87. Stires and Murphy, *PERT CPM*, p. 85.
88. Ibid., p. 98.
89. See Thomas A. Marschak, "Models, Rules of Thumb, and Development Decisions," in Dean, ed., *Operations Research*, p. 247. That this approach is normative rather than empirical is clearly indicated by his conclusion that such a manager "ought to re-examine," etc.
90. Bruce D. Henderson, in Bright, ed., *Technological Planning*, p. 247. Most of the papers in Dean, ed., *Operations Research,* carry this same message: "Only by continually improving decision procedures will management be able to meet the competition from companies that have improved their management decision procedures through Operations Research."—David B. Hertz and Phillip G. Carlson, "Selection, Evaluation, and Control of Research and Development Projects," in Dean, ed., *Operations Research*, p. 170.
91. Donald G. Malcolm, "Integrated Research and Development Management Systems," in Dean, ed., *Operations Research*, p. 131.
92. See Likert and Seashore, "Making Cost Control Work." Somewhat the same idea is contained in recent demands for a system of national social accounts. See, for example, Bertram Gross, "The Social State of the Nation," *Trans–Action*, vol. 3, no. 1 (November–December 1965).
93. Boguslaw, *New Utopians*; Rubenstein, in Dean, ed., *Operations Research*, p. 192, says: "Most of the early formulations for project selection and even those currently under development evade the issue of how the estimates on these formulations are derived."
94. Thomas B. Ross in the Chicago *Sun-Times* (June 25, 1967), p. 7.
95. The Chicago *Sun-Times* (June 18, 1967), section 2, p. 8.
96. I refer to the data-processing industry. Critical Path Method

(CPM) and Minimum Cost Expediting (MCX) were originated late in 1956 by a few people from Du Pont's Integrated Engineering Control Group, aided by a chap from Univac Applications Research Center. See William F. Ashley and Milton T. Austin, "Case Studies in Network Planning, Scheduling, and Control of Research and Development Projects," in Dean, ed., *Operations Research*, pp. 264–65.

97. Michel Crozier mentions the frequent use in French administration of the *grève du zèle*—a form of strike in which employees make a point of obeying all of management's rules to the letter, and by this means effect a significant reduction in work output. European workers have often resorted to this kind of "sabotage." See *The Bureaucratic Phenomena* (Chicago: University of Chicago Press, 1964), p. 139.

98. Wildavsky's study of budgeting (*The Politics of the Budgetary Process*) is an exception. In a rather casual survey of literature on research I have found nineteen adaptations designed to protect "controlees" from "controllers," including (among others): persistent optimism in estimates, selective forgetting, selective inattention, "bootlegging" (fund-juggling), purposely vague project proposals, differentiation and segregation of research activities from more pedestrian activities, pretended administrative ineptness, control of spokesmen (by rewarding scientist supervisors for their scientific, not their administrative, capacities), and adherence to the ideology of "pure science."

99. Jesse Burkhead, "The Budget and Democratic Government," in Roscoe C. Martin, ed., *Public Administration and Democracy* (Syracuse, N.Y.: Syracuse University Press, 1965), pp. 85–99.

100. Likert and Seashore ("Making Cost Control Work") cast doubt on the efficiency, even, of present production management methods, since they proceed without data as to the intervening variables, data concerning the internal health of the social organisms. See also Louis B. Barnes, "Organizational Systems and Engineering Groups," in Orth, Bailey and Wolek, *Administering Research and Development*, pp. 72–85.

101. "The stresses related to R and D management in Defense has thus resulted in a chaotic and degraded management situation with a serious reduction in R and D efficiency. Furthermore, this situation has had adverse effects on the performance of military R and D in industry."—Ellis A. Johnson, "A Proposal for Strengthening U.S. Technology," in Dean, ed., *Operations Research*, p. 12. Donald

Zylstra in the Syracuse *Post-Standard* says: "Concern with 'cost-effectiveness' has promoted unusual technical caution in the Pentagon. Prolonged studies and re-studies have delayed most technological advances in weapons." He goes on to say: "Only in the field of new weapons research and development has McNamara been unable to silence a growing group of critics."

102. The similarity of current federal administration and the early, Taylor-dominated administration of the Forest Service is startling. See Ashley L. Schiff, "Innovation and Administrative Decision Making: The Conservation of Land Resources," *Administrative Science Quarterly*, vol. 11, no. 1 (June 1966), pp. 1–30. In an "Initial Memorandum," the Subcommittee on National Security and International Affairs (90th Cong., 1st Sess.), stated that cost-benefit analysis goes back to the Garden of Eden (see Genesis 3), and that performance budgeting goes back at least to President Taft's Commission on Economy and Efficiency. To the same general effect see Frederick C. Mosher, "PPBS: Two Questions," *Public Administration Review*, vol. 27, no. 1 (March 1967). The old keeps popping up in new pretentious terminology.

CHAPTER IV

1. See Thomas S. Kuhn, *The Structure of Scientific Revolutions* (Chicago: University of Chicago Press, 1962).

2. James G. March, ed., *Handbook of Organizations* (Chicago: Rand McNally & Company, 1965).

3. Ibid., p. 423.

4. Ibid., p. 650.

5. Jacob Perlman, "Measurements of Scientific Research and Development," in Burton V. Dean, ed., *Operations Research in Research and Development* (New York: John Wiley & Sons, Inc. 1963), pp. 42–96.

6. Barkev S. Sanders, "Some Difficulties in Measuring Inventive Activities," in National Bureau of Economic Research, (NBER), *The Rate and Direction of Inventive Activities* (Princeton, N.J.: Princeton University Press, 1962), pp. 58–59.

7. National Science Foundation, *Reviews of Data on Research and Development*, no. 41 (September 1963) p. 1.

8. Sanders, "Some Difficulties," p. 60; Fritz Machlup, "The Sup-

ply of Inventors and Inventions," in NBER, *Rate and Direction*, pp. 143-67.

9. S. C. Gilfillan, comments in NBER, *Rate and Direction*, pp. 83-84.

10. See Sanders, "Some Difficulties," p. 58.

11. Gilfillan, *The Sociology of Invention* (Federalsburg, Md.: Stowell, 1935), p. 109.

12. *Statistical Abstract of the United States* for 1965, tables 2 and 768.

13. Francis Bello, "Industrial Laboratory," *Fortune*, November 1958, p. 214; Miles W. Martin, Jr., "The Measurement of Value of Scientific Information," in Dean, ed., *Operations Research*, pp. 97-123; Estelle Brodman, "Choosing Physiology Journals," *Medical Library Association Bulletin,* vol. 32 (1944), pp. 479-83.

14. Gilfillan, in NBER, *Rate and Direction*, p. 84.

15. From a mimeographed working document.

16. Chris Argyris, *Organization and Innovation* (Homewood, Ill.: Richard D. Irwin, Inc., 1965).

17. See W. O. Hagstrom, *The Scientific Community* (New York: Basic Books, 1965); Simon Marcson, *The Scientist in American Industry* (Princeton, N.J.: Industrial Relations School, 1960); William Kornhauser, *Scientists in Industry: Conflicts and Accommodation* (Berkeley: University of California Press, 1962); Opinion Research Corporation, *The Conflict Between the Management Mind and the Scientific Mind* (Princeton, N.J.: Opinion Res. Corp., 1959); Howard S. Becker and James Cooper, "The Development of Identification with an Occupation," *American Journal of Sociology*, vol. 61 (January 1956), pp. 289-98; Norman W. Storer, "Some Sociological Aspects of Federal Science Policy," *American Behavioral Scientist*, vol. 6 (December 1962), pp. 27-30.

18. I have observed that production-oriented people who stress purposiveness are also likely to stress the importance of "leadership." However, surely the rise of leadership is related to goal vagueness and ambiguity. Purpose and program are substitutes for leadership. This paradox is explained if it is true, as I think it is, by the fact that the managerially oriented theorists usually use "leadership" when they mean "headship" or "authority role." This use of the term serves the ideological function of helping to legitimatize the *status quo*.

CHAPTER V

1. See the growing literature in the field of comparative administration. Especially important is Fred W. Riggs, "Agraria and Industria—Toward a Typology of Comparative Administration," in William J. Siffin, ed., *Toward the Comparative Study of Public Administration* (Bloomington: Indiana University Press, 1959), pp. 23–116. See also Victor A. Thompson, "Bureaucracy in a Democratic Society," in Roscoe Martin, ed., *Public Administration and Democracy* (Syracuse, N.Y.: Syracuse University Press, 1965).

2. See Emile Durkheim, *The Division of Labor in Society*, trans. George Simpson (New York: The Macmillan Co., 1933); Eric Fromm, *Escape from Freedom* (New York: Holt, Rinehart & Winston, Inc., 1941); and Daniel Lerner, *The Passing of Traditional Society* (Glencoe, Ill.: The Free Press, 1958).

3. See, for example, Clark Kerr, John T. Dunlop, Frederick Harbison, and Charles A. Myers, *Industrialism and Industrial Man* (New York: Oxford University Press, 1964).

4. See Francois Russo, "Scientific and Technical Creation," *American Behavioral Scientist*, vol. 6, no. 5 (January 1961), pp. 33–34.

5. In 1962, 15.1 per cent of federal civilian personnel were professionals. In 1960, scientists and engineers constituted the following percentages of total personnel of the following agencies:

All	— 10.4	Agriculture	— 32.8	NASA	— 52.0
State	— 12.1	Commerce	— 24.8	TVA	— 39.2
Army	— 11.9	Labor	— 12.6	VA	— 25.5
Navy	— 15.7	HEW	— 19.4	Others	— 8.8
Air Force	— 8.2	AEC	— 19.7		
Interior	— 27.8	FAA	— 6.5		

6. A useful index of industrialization can be derived by dividing the highest civil service salary by the lowest. The resulting factor will be around 50 in very underdeveloped countries and around 5 in highly industrialized ones.

7. See especially William Kornhauser, *Scientists in Industry: Conflicts and Accommodation* (Berkeley: University of California Press, 1962), ch. 4. Recent research indicates that most workers, not just pro-

fessionals, are motivated chiefly by factors related to the work itself rather than the more conventional rewards administrated by managements. In other words, the intrinsic rewards of work are more important to most people than the extrinsic rewards of power, status, and income. See Frederick Herzberg, Bernard Mausner, and Barbara Snyderman, *The Motivation to Work* (New York: John Wiley & Sons, Inc., 1959); and M. Scott Meyers, "The Management of the Motivation to Work" (unpublished report on a motivation research project at Texas Instrument Company).

8. For an exposition of this important concept, see Gary S. Becker, *Human Capital* (New York: Columbia University Press, 1964). The notion of labor as a commodity is inextricably bound up with management's perennial attempt to get more work for less pay and its necessarily associated belief that all workers want is more pay for less work.

9. O. Glenn Stahl of the U.S. Civil Service Commission, "Functional Classification System for Scientists and Engineers," *Public Administration Review*, vol. 26, no. 4 (December 1966), p. 356.

10. "It is . . . not an integration of the humanistic and natural sciences cultures of which [we] stand in need but an integration of the professional and the civil culture." Edward Shils, "Demagogues and Cadres in the Political Development of the New States," in Lucien Pye, ed., *Communication and Political Development* (Princeton, N.J.: Princeton University Press, 1963), p. 76. "The central problem facing society with regard to all of the professions is the same: how can they be controlled without having their effectiveness destroyed?" Warren O. Hagstrom, *The Scientific Community* (New York: Basic Books, 1965), p. 294.

11. The following from a newspaper story illustrates the process rather forcefully: "The prototype of the federal recreation worker is the national park guide or the forest ranger. But state and federal governments hire recreation workers and supervisors in 60 categories, including biologists, administration personnel, enforcement officers, architects, nurses, historians, psychologists, chemists, geologists, therapists, economists, archeologists, as well as those one expects—tree experts, naturalists, fish hatchery managers, rangers and craft teachers." D. J. R. Bruckner, "Recreation Field Boom Producing Many Jobs," Syracuse *Post-Standard*, June 24, 1965, p. 4.

12. An examination of 16 engineering union contracts found

the following items included: (1) employer payment of professional society dues, (2) paid time-off to attend professional society meetings, (3) tuition refunds, (4) leaves of absence for educational reasons, and (5) maintenance of standards on the job. Kornhauser, *Scientists in Industry*, p. 104. A survey of 73 National Industrial Conference Board companies showed 56 paid employees' dues in professional societies. Ibid., p. 88. See also William Mussman, "Subsidizing Membership in Professional and Technical Societies," *Management Record*, vol. 13, no. 4 (April 1951), pp. 140–41.

13. "Some companies, indeed, are already faced with the necessity of paying a bright young mathematician fresh out of graduate school a salary equivalent to that being drawn by a manager with fifteen to twenty years service."—Samuel E. Hill and Frederick Harbison, *Manpower and Innovation in American Industry* (Princeton, N.J.: Industrial Relations Section, Department of Economics and Sociology, Princeton University, 1959), p. 63.

14. See Robert Presthus, *The Organizational Society* (New York: Alfred A. Knopf, 1962); and Aaron Levenstein, *Why People Work: Changing Incentives in a Troubled World* (New York: Crowell-Collier Press, 1962). "At college after college an increasing percentage of graduates is shunning business careers in favor of such fields as teaching, scientific research, law and public service . . . . 'We are deeply concerned with the number of college youths who have rejected business as a career,' says John E. Harmon, director of manpower development and training at the U.S. Chamber of Commerce. 'We're worried about the poor attitude of many students toward business.' . . . Comments Robert W. Feagles, personnel vice president of First National City Bank of New York: 'It's harder to get good men, even though there are more college graduates than ever.' "—Roger Ricklefs in *The Wall Street Journal* (Nov. 10, 1964). On the growing motivational problem among professionals, see ch. 3, note 52. Much of the current motivational problem is due to administrative obsolescence, discussed below.

15. See his *The New Science of Management Decision* (New York: Harper & Brothers Publishers, 1960).

16. James D. Thompson and Arthur Tuden, "Strategies, Structures, and Processes of Organizational Decision," in *Comparative Studies in Administration* (Pittsburgh: University of Pittsburgh Press, 1959).

17. Because he has been such an eloquent and learned spokesman for the "New Science" I refer again to Herbert Simon, who argues that the promise of the "New Science" is not economy but control of a large number of variables.—*New Science of Management Decision*, p. 45.

18. Chester Barnard, *The Functions of the Executive* (Cambridge, Mass.: Harvard University Press, 1938). "We pay large inventory costs, also, to permit factory and sales managements to make decisions in semi-independence of each other. . . . With the development of operations research . . . and . . . the technical means to maintain and adjust the data . . . large savings are attainable . . . ."—Simon, *New Science of Management*, pp. 45–46.

19. Note, however, the following comments of an air force officer on the effects of the SAGE application (a highly automated control system in the North American Defense Command): "One of the queerest observations that I have made concerns this mass of engineers, technicians, machine operators, and operations people milling around and working almost unaware that anyone else exists. That is to say there doesn't seem to be any interaction between the individuals. All of the interaction seems to be with the electronic system. This is quite a change from the old squadron where communication and interaction between individuals were a must to accomplish the mission. In addition, it carried over into the social environment and it developed friendships, cliques and competitions. This leads to a question about the importance of tradition, regulations, lines of authority, and morale. These have always been an integral part of military organizations, but, in this instance, they seem to be relatively unimportant. I believe that the computer is the cohesive element in these up and coming systems, and they simply set the pace and individuals blindly follow. It's like a fire into which everyone is throwing everything he owns for fear that, should it go out, they will all die of the cold."—Thomas L. Whisler, "Executives and Their Jobs—the Changing Organizational Structure," Selected Papers No. 9 (Chicago: Graduate School of Business, University of Chicago, n.d.), pp. 6–7. See also Alan F. Westin, *Privacy and Freedom* (New York: Atheneum, 1967), especially chapter 7.

20. This point is made by a number of contributors to National Bureau of Economic Research, *The Rate and Direction of Inventive Activities* (Princeton, N.J.: Princeton University Press, 1962). For example, the pieces by Jesse W. Markham, Jora R. Minasian, and

Charles J. Hitch. See also U.S. Bureau of the Budget, *Measuring Productivity of Federal Government Organizations* (Washington: G.P.O., 1964), pp. 9–10; Joseph A. Schumpeter, *Capitalism, Socialism and Democracy* (New York: Harper & Brothers, 1950), especially ch. 7; James R. Bright, ed., *Technological Planning on the Corporate Level* (Cambridge, Mass.: Harvard University Graduate School of Business, 1962), p. 96; and Robert M. Solow, "Technological Change and the Aggregate Production Function," *Review of Economics and Statistics*, vol. 39 (August 1957), pp. 312–20.

21. Francis Bello, "The Technology Behind Productivity," in John T. Dunlop, ed., *Automation and Technological Change* (Englewood Cliffs, N.J.: Prentice-Hall, Inc., 1962), pp. 159–63; Richard N. Cooper, "International Aspects," in Dunlop, ed. *Automation and Technological Change,* pp. 132–52.

22. See especially Bertram M. Gross, *The State of the Nation: Social Systems Accounting* (London: Tavistock Publications, 1966). Although the author describes this work as "studiously resisting the temptation to leap precipitately into premature use of the proposed accounting system for the purpose of prediction and control," his managerial orientation is apparent throughout the book. For example, he says that "the great value of social systems accounting" is that it provides data from which someone can derive "criteria for evaluating past and present situations and trends [and] strategic objectives for current action . . ." (p. 144). I do not want to neglect the enormous importance to science and technology of the computational aspects of the new data-processing technology.

23. For a journalistic treatment of secrecy in the federal government during the 1955–1965 decade, see Miles Beardsley Johnson, *The Government Secrecy Controversy* (New York: Vantage Press, Inc., 1967). I have discussed the functional bases of modern administrative secrecy in "Bureaucracy in a Democratic Society."

24. The following analysis is taken from Hagstrom, *The Scientific Community*, pp. 19–20. On the incompatibility of traditional management practices and professionalism, see Simon Marcson, *The Scientist in American Industry* (Princeton, N.J.: Industrial Relations School, 1960), especially p. 136. See also Kornhauser, *Scientists in Industry*.

25. See Norman W. Storer, "Some Sociological Aspects of Federal Science Policy," *A.B.S.,* vol. 6 (December 1962). He suggests small

specialized enclaves within the scientific community based on a greater proliferation of fields of interest, new journals, etc., all to help retain the meaningfulness of professional recognition as a reward and hence a control.

26. Herzberg, Mausner, and Snyderman, *The Motivation to Work*. Others have obtained results similar to theirs.

Bibliography

Alexander-Frutschi, M. C. *An Annotated Bibliography on Small Industry Development.* Glencoe, Ill.: The Free Press, 1959.
———. *Human Resources and Economic Growth.* Menlo Park, Calif.: Stanford Research Institute, 1963.
Allen, G. H., ed. *Individual Initiative in Business.* Cambridge: Harvard University Press, 1950.
Ambrose, Stephan E., ed. *Institutions in Modern America.* Baltimore: The Johns Hopkins Press, 1967.
Anderson, H. H., ed. *Creativity and Its Cultivation.* New York: Harper & Row, 1959.
Anderson, N. *Work and Leisure.* New York: The Free Press, 1961.
———. *Dimensions of Work; the Sociology of a Work Culture.* New York: David McKay Co., Inc., 1964.
Argyris, Chris. *Organization and Innovation.* Homewood, Ill.: Richard D. Irwin, Inc., 1965.
———. *Personality and Organization: the Conflict between System and the Individual.* New York: Harper & Row, 1957.
Armytage, W. H. G. *A Social History of Engineering.* New York: Pitman Publishing Corp., 1961.
Arrow, Kenneth J., "Economic Welfare and the Allocation of Resources for Invention." In National Bureau of Economic Research, *The Rate and Direction of Inventive Activities.* Princeton, N.J.: Princeton University Press, 1962.
Ashley, William F., and Milton T. Austin. "Case Studies in Network Planning, Scheduling, and Control of Research and Development

Projects." In *Operations Research in Research and Development,* edited by Burton V. Dean. New York: John Wiley & Sons, Inc., 1963.

Back, Kurt W. "Decisions under Uncertainty," *The American Behavioral Scientist,* vol. 4, no. 6 (February 1961).

Baldwin, W. L. *Contracted R and D for Small Business.* Hanover, N.H.: Amos Tuck School of Business Administration, Dartmouth College, 1962.

Barnes, Louis B. "Organizational Systems and Engineering Groups." In *Administering Research and Development: The Behavior of Scientists and Engineers in Organization,* edited by Charles D. Orth, III, Joseph C. Bailey, and Francis Wolek. Homewood, Ill.: Richard D. Irwin, Inc., 1964.

Barnett, H. G. *Innovation: A Basis for Cultural Change.* New York: McGraw-Hill Book Co., 1953.

Baumol, William J. *Economic Theory and Operations Analysis.* Englewood Cliffs, N.J.: Prentice-Hall, Inc., 1961.

Beaumont, R., ed. *Applying Behavioral Science Research to Industry.* New York: Industrial Relations Counselors, 1964.

Beaumont, Richard, and Roy B. Helfgott. *Management, Automation, and People.* New York: Industrial Relations Counselors, 1964.

Becker, Gary S. *Human Capital.* New York: Columbia University Press, 1964.

Becker, Howard S., and J. W. Carper. "The Development of Identification with an Occupation," *American Journal of Sociology,* vol. 61 (January 1956), pp. 289–98.

Bello, Francis. "Industrial Laboratory," *Fortune,* November 1958.

———. "The Technology Behind Productivity." In *Automation and Technological Change,* edited by John T. Dunlop. Englewood Cliffs, N.J.: Prentice-Hall, Inc., 1962.

Bennis, W. G., Kenneth D. Berne, and Robert Chin. *The Planning of Change.* New York: Holt, Rinehart, & Winston, 1962.

Berelson, B., and G. A. Steiner. *Human Behavior: An Inventory of Scientific Finding.* New York: Harcourt Brace & World, 1964.

Berle, A. K., and L. S. DeCamp. *Inventions and Their Management.* Scranton, Pa.: International Textbook Co., 1951.

Beveridge, W. I. B. *The Art of Scientific Investigation.* New York: Vintage Books, 1950.

Biderman, A. D., and H. Zimmer. *The Manipulation of Human Behavior.* New York: John Wiley & Sons, Inc., 1961.

Blau, Peter M. *The Dynamics of Bureaucracy*, rev. ed. Chicago: University of Chicago Press, 1963.
Blau, Peter M., and W. Richard Scott. *Formal Organizations*. San Francisco: Chandler Publishing Co., 1962.
Blauner, R. *Alienation and Freedom: The Factory Worker and His Industry*. Chicago: University of Chicago Press, 1964.
Bloomberg, W., Jr. *The Age of Automation*. New York: League for Industrial Democracy, 1955.
Boguslaw, Robert. *The New Utopians: A Study of System Design and Social Change*. Englewood Cliffs, N.J.: Prentice-Hall, Inc., 1965.
Boulding, K. E. *The Organizational Revolution*. New York: Harper & Row, 1953.
Brady, R. A. *Organization, Automation and Society: The Scientific Revolution in Industry*. Berkeley: University of California Press, 1961.
Braybrooke, David, and Charles E. Lindblom. *A Strategy of Decision*. New York: The Macmillan Co., 1963.
Bright, James R. *Automation and Management*. Boston: Graduate School of Business Administration, Harvard University, 1958.
―――. *Research, Development, and Technological Innovation*. Homewood, Ill.: Richard D. Irwin, Inc., 1964.
―――, ed. *Technological Planning on the Corporate Level*. Cambridge: Harvard University Graduate School of Business, 1962.
Brodman, Estelle. "Choosing Physiology Journals," *Medical Library Association Bulletin*, vol. 32, no. 4 (October 1944), pp. 479–83.
Bross, Irwin D. J. *Design for Decision*. New York: The Macmillan Co., 1953.
Brown, J. Douglas, and Frederick Harbison. *High-Talent Manpower for Science and Industry*. Princeton: Princeton University Press, 1957.
Buckingham, W. S. *Automation: Its Impact on Business and People*. New York: Harper & Row, 1965.
Buffa, Elwood S. *Modern Production Management*, rev. ed. New York: John Wiley and Sons, Inc., 1965.
Burkhead, Jesse. "The Budget and Democratic Government." In *Public Administration and Democracy*, edited by Roscoe C. Martin. Syracuse, N.Y.: Syracuse University Press, 1965.
Burns, Tom, and B. M. Stalker. *The Management of Innovation*. London: Tavistock Publications, 1959.

Caplow, Theodore. *The Sociology of Work*. Minneapolis: University of Minnesota Press, 1954.
Caplow, Theodore, and Reece J. McGee. *The Academic Marketplace*. New York: Basic Books, 1958.
Cartwright, Dorwin, and Alvin Zander, eds. *Group Dynamics*. 2nd ed. Evanston, Ill.: Row, Peterson & Co., 1962.
Chapple, Eliot D., and Leonard R. Sayles. *The Measure of Management*. New York: The Macmillan Co., 1961.
Cheit, E., ed. *The Business Establishment*. New York: John Wiley & Sons, Inc., 1964.
Clark, C. A. *Brainstorming*. New York: Doubleday & Co., Inc., 1958.
Clark, J. H. *Competition as a Dynamic Process*. Washington: Brookings Institution, 1961.
Cochran, T. C. *Railroad Leaders, 1845–1890; the Business Mind in Action*. Cambridge: Harvard University Press, 1953.
Cohen, Harry. *The Demonics of Bureaucracy*. Ames, Iowa: The Iowa State University Press, 1965.
Collier, A. T. *Management Men and Values*. New York: Harper & Row, 1962.
Comparative Studies in Administration. Edited by the staff of the Administrative Sciences Center. Pittsburgh: University of Pittsburgh Press, 1959.
Cooper, Richard N. "International Aspects." In *Automation and Technological Change,* edited by John T. Dunlop. Englewood Cliffs, N.J.: Prentice-Hall, Inc., 1962.
Corsini, R. J., et al. *Role Playing in Business and Industry*. Glencoe, Ill.: The Free Press, 1961.
Coser, L. *The Functions of Social Conflict*. Glencoe, Ill.: The Free Press, 1956.
Croome, Honor. *Human Problems of Innovation*. London: H.M. Stationery Office, 1960.
Crozier, Michel. *The Bureaucratic Phenomena*. Chicago: University of Chicago Press, 1964.
Cyert, Richard M., and James G. March. *A Behavioral Theory of the Firm*. Englewood Cliffs, N.J.: Prentice-Hall, Inc., 1963.
Cyert, Richard M., James G. March, and William A. Starbuck. "Two Experiments on Bias and Conflict in Organizational Estimation," *Management Science*, vol. 7, no. 3 (April 1961).
Cyert, Richard M., W. R. Dill, and J. G. March. "The Role of Ex-

pectations in Business Decision-Making," *Administrative Science Quarterly*, vol. 3, no. 3 (December 1958).

Dalton, Malville. "Conflicts between Staff and Line Mangerial Officers," *American Sociological Review*, vol. 15 (June 1950), pp. 342–51.

———. *Men Who Manage*. New York: John Wiley & Sons, Inc., 1960.

Danieve, A. *Higher Education in the American Economy*. New York: Random House, Inc., 1964.

David, H. *Manpower Policies for a Democratic Society*. New York: Columbia University Press (for National Manpower Council), 1965.

———. *Public Policies and Manpower Resources*. New York: Columbia University Press, 1964.

———. *Education and Manpower*. New York: Columbia University Press, 1960.

Dean, Burton V., ed. *Operations Research in Research and Development*. New York: John Wiley & Sons, Inc., 1963.

De Grazia, S. *Of Time, Work and Leisure*. New York: Twentieth Century Fund, 1962.

Diebold, J. *Beyond Automation: Managerial Problems of Exploding Technology*. New York: McGraw-Hill Book Co., 1964.

Dror, Yehezkel. *Public Policy Making Reexamined*. San Francisco: Chandler Publishing Co., 1968.

Dunlop, John T., ed., *Automation and Technological Change*. Englewood Cliffs, N.J.: Prentice-Hall, Inc., 1962.

Dunnette, M. "Are Meetings Any Good for Solving Problems?" *Personnel Administration*, vol. 27 (March–April 1964).

Durkheim, Emile. *The Division of Labor in Society*. Translated by George Simpson. New York: The Macmillan Co., 1933.

Endicott, Frank A. *Trends in Employment of College and University Graduates in Business and Industry*. Evanston: Northwestern University Press, 1962.

Enos, John L. "Invention and Innovation in the Petroleum Industry." In National Bureau of Economic Research, *The Rate and Direction of Inventive Activities*. Princeton, N.J.: Princeton University Press, 1962.

Enthoven, Alain, and Harry Rowen. "Defense Planning and Organization." In National Bureau of Economic Research, *Public Fi-*

nances, Needs, Sources, and Utilizations. Princeton, N.J.: Princeton University Press, 1961.

Etzione, A. *A Comparative Analysis of Complex Organizations*. New York: The Free Press, 1961.

———. *Modern Organizations*. Englewood Cliffs, N.J.: Prentice-Hall, Inc., 1964.

———, ed. *Social Change: Sources, Patterns and Consequences*. New York: Basic Books, Inc., 1964.

Fellner, William. "Does the Market Direct the Relative Factor-Saving Effects of Technological Progress?" In National Bureau of Economic Research, *The Rate and Direction of Inventive Activities*. Princeton, N.J.: Princeton University Press, 1962.

Friedman, G. *The Anatomy of Work: Labor, Leisure and the Implications of Automation*. Glencoe, Ill.: The Free Press, 1961.

Fromm, Eric. *Escape from Freedom*. New York: Holt, Rinehart & Winston, Inc., 1941

Galbraith, John Kenneth. *American Capitalism*. Boston: Houghton Mifflin Co., 1952.

Gantt, Henry. *Industrial Leadership*. New Haven: Yale University Press, 1916.

Gerard, R. W. *Mirror to Physiology: A Self-Survey of Physiological Science*. Washington: American Physiological Society, 1958.

Ghiselin, B. *The Creative Process; a Symposium*. New York: New American Library, Inc., 1955.

Gibb, Cecil A. "Leadership." In *Handbook of Social Psychology*, edited by Gardner Lindsey. Reading, Mass.: Addison-Wesley Publishing Co., Inc., 1954.

Gilfillan, S. C. *The Sociology of Invention*. Federalsburg, Md.: Stowell, 1935.

Ginzberg, E., and H. Berman. *The American Worker in the Twentieth Century. A History Through Autobiography*. New York: The Free Press, 1963.

Ginzberg, E., and J. L. Hernia, et al. *Talent and Performance*. New York: Columbia University Press, 1964.

———. *Technology and Social Change*. New York: Columbia University Press, 1964.

Ginzberg, E., and E. W. Reilly. *Effecting Change in Large Organizations*. New York: Columbia University Press, 1957.

———. *Human Resources, The Wealth of a Nation*. New York: Simon & Schuster, Inc., 1958.

Glasner, Barney G. *Organizational Scientists: Their Professional Careers.* Indianapolis: Bobbs-Merrill Co., 1964.

Golembiewski, R. T. *The Small Group: An Analysis of Research Concepts and Operations.* Chicago: University of Chicago Press, 1962.

Gordon, Gerald, and Selwyn Becker. "Changes in Medical Practice Bring Shifts in the Pattern of Power," *The Modern Hospital,* vol. 102, no. 2 (February 1964).

Gordon, W. J. J. *Synectics: The Development of Creative Capacity.* New York: Harper & Row, 1961.

Gouldner, Alvin W. "Cosmopolitans and Locals," *Administrative Science Quarterly,* vol. 2, no. 3 (December 1957), pp. 281–306, and no. 4 (March 1958), pp. 444–80.

Gouldner, Alvin W. *Patterns of Industrial Bureaucracy.* Glencoe, Ill.: The Free Press, 1954.

Gross, Bertram M. "The Social State of the Nation," *Trans-Action,* vol. 3, no. 1 (November–December 1965).

———. *The State of the Nation: Social Systems Accounting.* London: Tavistock Publications, 1966.

Gruber, H. E., and G. Terrell, *et al.,* eds. *Contemporary Approaches to Creative Thinking.* New York: Atherton Press, 1962.

Guest, R. H. *Organizational Change: The Effects of Success on Leadership.* Homewood, Ill.: Richard D. Irwin, Inc., 1961.

Haefele, John W. *Creativity and Innovation.* New York: Reinhold Publishing Corp., 1962.

Hagen, Everett. *On the Theory of Social Change: How Economic Growth Begins.* Homewood, Ill.: Richard D. Irwin, Inc., 1962.

Hagstrom, W. O. *The Scientific Community.* New York: Basic Books, 1965.

Haire, M., ed. *Modern Organization Theory.* New York: John Wiley & Sons, Inc., 1959.

———, ed. *Organization Theory in Industrial Practice: A Symposium* . . . New York: John Wiley & Sons, Inc., 1962.

Harbison, F., and C. A. Myers, eds. *Manpower and Education: Country Studies in Economic Development.* New York: McGraw-Hill Book Co., 1965.

———, eds. *Education, Manpower, and Economic Growth: Strategies of Human Resources Development.* New York: McGraw-Hill Book Co., 1964.

———. *Management in the Industrial World: An International Analysis.* New York: McGraw-Hill Book Co., 1959.

Harbrecht, P. P., and A. A. Berle, Jr. *Towards the Paraproprietal Society*. New York: Twentieth Century Fund, 1960.
Harrington, Joseph, Jr., "A Look into Tomorrow." In *New Views on Automation*. Joint Economic Committee, 86th Congress, 2nd Session.
Harris, John S. "New Product Profile Chart." In James R. Bright, ed., *Research, Development, and Technological Innovation*. Homewood, Ill.: Richard D. Irwin, Inc., 1964.
Hartman, H. *Authority and Organization in German Management*. Princeton, N.J.: Princeton University Press, 1959.
Hatford, H. S. *The Inventor and His World*. London: Penguin Books, Ltd., 1948.
Henderson, Bruce D. "Implications of Technology for Management." In *Technological Planning on the Corporate Level*, edited by James R. Bright. Cambridge: Harvard University Graduate School of Business, 1962.
Hertz, David B., and Phillip G. Carlson. "Selection, Evaluation, and Control of Research and Development Projects." In *Operations Research in Research and Development*, edited by Burton V. Dean. New York: John Wiley & Sons, Inc., 1963.
Herzberg, Frederick, Bernard Mausner, and Barbara Snyderman. *The Motivation to Work*. New York: John Wiley & Sons, Inc., 1959.
Heyel, Carl, ed. *The Encyclopedia of Management*. New York: Reinhold Publishing Corp., 1963.
———. *Handbook of Industrial Research Management*. New York: Reinhold Publishing Corp., 1959.
Highet, G. *Talents and Geniuses*. New York: Oxford University Press, 1957.
Hilgard, E. R. "Creativity and Problem Solving." In *Creativity and Its Cultivation*, edited by H. H. Anderson. New York: Harper & Row, 1959.
Hill, Karl, ed. *The Management of Scientists*. Boston: Beacon Press, 1964.
Hill, Samuel E., and Frederick Harbison. *Manpower and Innovation in American Industry*. Princeton, N.J.: Industrial Relations Section, Department of Economics and Sociology, Princeton University, 1959.
Hirshman, Albert O. *The Strategy of Economic Development*. New Haven, Conn.: Yale University Press, 1958.

———, and Charles E. Lindblom. "Economic Development, Research and Development, Policy Making: Some Converging Views," *Behavioral Science*, vol. 7, no. 2 (April 1962), pp. 211–22.

Homans, G. *The Human Group*. New York: Harcourt Brace & World, Inc., 1950.

Hoos, I. R. *Automation in the Office*. Washington: Public Affairs Press, 1961.

How to Attract and Hold Engineering Talent. Executive Research Survey No. 3. Washington: Professional Engineers Conference Board for Industry, 1954.

Hower, R. M., and C. D. Orth. *Managers and Scientists: Some Human Problems in Industrial Research Organizations*. Boston: Harvard University Press, 1963.

Hughes, Everett C. *Men and Their Work*. Glencoe, Ill.: The Free Press, 1958.

James, William. *Psychology: Briefer Course*. New York: Henry Holt, 1923.

Janowitz, Morris. *The Professional Soldier*. Glencoe, Ill.: The Free Press, 1960.

Jewkes, J., et al. *The Sources of Invention*. London: Macmillan Co., 1958.

Johnson, Ellis A. "A Proposal for Strengthening U. S. Technology." In *Operations Research in Research and Development*, edited by Burton V. Dean. New York: John Wiley & Sons, Inc., 1963.

Johnson, Miles Beardsley. *The Government Secrecy Controversy*. New York: Vantage Press, Inc., 1967.

Kaempffort, Waldemar. "Systematic Invention," *Forum*, vol. 70, no. 4 (October 1923), pp. 2,010–18.

———. "Invention by Wholesale," *Forum*, vol. 70, no. 5 (November 1923), pp. 2,116–22.

———. *Invention and Society*. Chicago: American Library Association, 1930.

———, ed. *A Popular History of American Invention*. New York: Charles Scribner's Sons, 1924.

Kahn, R., et al. *Organization Stress: Studies in Role-Conflict and Ambiguity*. New York: John Wiley & Sons, Inc., 1964.

Kahn, R. L., and K. Boulding, eds. *Power and Conflict in Organization*. New York: Basic Books, 1964.

Kaplan, Norman. "Some Organizational Factors Affecting Creativity."

In *Administering Research and Development: The Behavior of Scientists and Engineers in Organization,* edited by Charles D. Orth, III, Joseph C. Bailey, and Francis Wolek. Homewood, Ill.: Richard D. Irwin, Inc., 1964.

Katz, Elihu. "The Social Itinerary of Technical Change: Two Studies of the Diffusion of Innovation." In *Studies of Innovation and of Communication to the Public.* Studies in the Utilization of Behavioral Science. Vol. 2. Stanford, Calif.: Institute for Communication Research, Stanford University, 1962.

Kerr, Clark, John T. Dunlop, Frederick Harbison, and Charles A. Myers. *Industrialism and Industrial Man.* New York: Oxford University Press, 1964.

Klein, Burton H. "A Radical Proposal for R. and D.," *Fortune,* vol. 57, (May 1958).

———. "The Decision Making Problem in Development." In National Bureau for Economic Research, *The Rate and Direction of Inventive Activities.* Princeton, N.J.: Princeton University Press, 1962.

———, and W. Meckling. "Application of Operations Research to Development Decisions," *Operations Research,* vol. 6, no. 3 (May–June 1958), pp. 352–63.

Kornhauser, William. *Scientists in Industry: Conflicts and Accommodation.* Berkeley: University of California Press, 1962.

Kuhn, Thomas S. *The Structure of Scientific Revolutions.* Chicago: University of Chicago Press, 1962.

LaPalombara, Joseph, ed. *Bureaucracy and Political Development.* Princeton, N.J.: Princeton University Press, 1963.

La Piere, R. T. *Social Change.* New York: McGraw-Hill Book Co., 1965.

Lapp, R. E. *The New Priesthood: The Scientific Elite and the Uses of Power.* New York: Harper & Row, 1965.

Large, Irving, David Fox, Joel Davitz, and Marlin Brenner. "A Survey of Studies Contrasting the Quality of Group Performance and Individual Performance, 1920–1957," *Psychological Bulletin,* vol. 55, no. 6 (November 1958), pp. 337–72.

Lerner, Daniel. *The Passing of Traditional Society.* Glencoe, Ill.: The Free Press, 1958.

———. "Toward a Communication Theory of Modernization." In *Communication and Political Development,* edited by Lucien Pye. Princeton, N.J.: Princeton University Press, 1963.

Levenstein, Aaron. *Why People Work: Changing Incentives in a Troubled World.* New York: Crowell-Collier Press, 1962.

Likert, Rensis. "Motivation and Increased Productivity," *Management Record,* vol. 18, no. 4 (April 1956).

———, and Stanley E. Seashore. "Making Cost Control Work," *Harvard Business Review,* vol. 41, no. 6 (November–December 1963), pp. 96–108.

Likert, Rensis. *New Patterns of Management,* New York: McGraw-Hill Book Co., 1961.

Lilley, C. *Automation and Social Progress.* New York: International Publishers, 1957.

Lindblom, Charles E. *The Intelligence of Democracy.* New York: The Free Press, 1965.

Lindsey, Gardner, ed. *Handbook of Social Psychology.* Reading, Mass.: Addison-Wesley Publishing Co., Inc., 1954.

Lippitt, Ronald, Jeanne Watson, and Bruce Westley. *The Dynamics of Planned Change.* New York: Harcourt Brace & World, Inc., 1958.

Lipset, S. M., and R. Bendix. *Social Mobility in Industrial Society.* Berkeley: University of California Press, 1959.

Lyden, Fremont J. "Brainstorming and Group Problem-Solving: The Same Thing?" *Public Administration Review,* vol. 25, no. 4 (December 1965).

———, and Ernest G. Miller, eds. *Planning, Programming, Budgeting: A Systems Approach to Management.* Chicago: Markheim Publishing Co., 1967.

Lynn, Kenneth S., and the editors of Daedulus, eds. *The Professions in America.* Boston: Beacon Press, 1967.

McClelland, David C. *The Achieving Society.* Princeton, N.J.: D. Van Nostrand Company, Inc., 1961.

Machlup, Fritz. "The Supply of Inventors and Inventions." In National Bureau for Economic Research, *The Rate and Direction of Inventive Activities.* Princeton, N.J.: Princeton University Press, 1962.

Maier, N. R. F. *Problem Solving Discussions and Conferences.* New York: McGraw-Hill Book Co., 1963.

———, and J. J. Hayes. *Creative Management.* New York: John Wiley & Sons, Inc., 1962.

Mainzer, Lewis C. "The Scientist as Public Administrator," *The*

Western Political Quarterly, vol. 16, no. 4 (December 1963), pp. 814–29.

Malcolm, Donald G. "Integrated Research and Development Management Systems." In *Operations Research in Research and Development,* edited by Burton V. Dean. New York: John Wiley & Sons, Inc., 1963.

Mannheim, Karl. *Ideology and Utopia.* New York: Harcourt, Brace, & Co., 1936.

Mansfield, E. "Technical Change and the Rate of Imitation," *Econometrica,* vol. 29 (October 1961), pp. 741–66.

March, James G., ed. *Handbook of Organizations.* Chicago: Rand McNally & Co., 1965.

———, and Herbert A. Simon. *Organizations.* New York: John Wiley & Sons, Inc., 1958.

Marcson, Simon. *The Scientist in American Industry.* Princeton, N.J.: Industrial Relations School, 1960.

Marschak, Thomas A. "Models, Rules of Thumb, and Development Decisions." In *Operations Research in Research and Development,* edited by Burton V. Dean. New York: John Wiley & Sons, Inc., 1963.

Marshall, A. W., and W. H. Meckling. "Predictability of the Cost, Time, and Success of Development." In National Bureau of Economic Research, *The Rate and Direction of Inventive Activities.* Princeton, N.J.: Princeton University Press, 1962.

Martin, Miles W., Jr. "The Measurement of Value of Scientific Information." In *Operations Research in Research and Development,* edited by Burton V. Dean. New York: John Wiley & Sons, Inc., 1963.

Martin, Roscoe C., ed. *Public Administration and Democracy.* Syracuse, N.Y.: Syracuse University Press, 1965.

Maslow, A. H. *Motivation and Personality.* New York: Harper & Brothers, 1954.

Masserman, J. H., ed. *Communications and Community.* New York: Grune & Stratton, Inc., 1965.

May, Rollo. *The Meaning of Anxiety.* New York: The Ronald Press Co., 1950.

Meier, R. L. *A Communication Theory of Urban Growth.* Cambridge: The Massachusetts Institute of Technology Press, 1962.

Meyers, M. Scott. "The Management of Motivation to Work." Un-

published report on a motivation research project at Texas Instrument Company.
Miles, Mathew B., ed. *Innovation in Education.* New York: Columbia University Teacher's College Press, 1964.
Miller, William, ed. *Men in Business.* Cambridge: Harvard University Press, 1952.
Millis, Walter. *Arms and Men: A Study in American Military History.* New York: G. P. Putnam's Sons, 1956.
Moore, David G., and Richard Renck. "The Professional in Industry." In *Administering Research and Development: The Behavior of Scientists and Engineers in Organization,* edited by Charles D. Orth, III, Joseph C. Bailey, and Francis Wolek. Homewood, Ill.: Richard D. Irwin, Inc., 1964.
Moore, W. E. *Social Change.* Englewood Cliffs, N.J.: Prentice-Hall, Inc., 1963.
Mueller, R. E. *Inventivity: How Man Creates in Art and Science.* New York: John Day Co., Inc., 1963.
Mueller, Willard F. "The Origins of the Basic Inventions Underlying DuPont's Major Product and Process Innovations, 1920 to 1950." In National Bureau of Economic Research, *The Rate and Direction of Inventive Activities.* Princeton, N.J.: Princeton University Press, 1962.
Municipal Manpower Commission. *Governmental Manpower for Tomorrow's Cities.* New York: McGraw-Hill Book Co., Inc., 1962.
Mussman, William. "Subsidizing Membership in Professional and Technical Societies," *Management Record,* vol. 13, no. 4 (April 1951).
National Academy of Science, Committee on Science and Public Policy. *Federal Support of Basic Research in Institutions of Higher Learning.* Washington, D.C.,: Government Printing Office, 1964.
National Bureau for Economic Research. *The Rate and Direction of Inventive Activities.* Princeton, N.J.: Princeton University Press, 1962.
———. *Public Finances, Needs, Sources, and Utilizations.* Princeton, N.J.: Princeton University Press, 1961.
National Manpower Council. *Government and Manpower.* New York: Columbia University Press, 1964.
National Science Foundation. Most NSF publications are relevant to the subject of this book.

Nelson, Richard R. "The Economics of Parallel R and D Efforts." Santa Monica: Rand Corp., November 1959.

———. "The Economics of Invention: A Survey of the Literature," *Journal of Business,* vol. 32, no. 2 (April 1959).

———. "The Simple Economics of Basic Scientific Research," *Journal of Political Economy,* vol. 67, no. 3 (June 1959).

New Views on Automation. Joint Economic Committee, 86th Congress, 2nd Session. Washington, D.C.: G.P.O., 1960.

Nord, O. C. *Growth of a New Product: Effect of Capacity in Acquisition Process.* Cambridge: The Massachusetts Institute of Technology Press, 1963.

Nosow, S., and W. H. Form, eds. *Man, Work and Society: A Reader in the Sociology of Occupations.* New York: Basic Books, 1962.

Novick, David, ed. *Program Budgeting.* Washington: Government Printing Office, 1965.

O'Brien, M. P. "Technological Planning and Misplanning." In *Technological Planning on the Corporate Level,* edited by James R. Bright. Cambridge, Mass.: Harvard University Graduate School of Business, 1962.

Ogburn, W. F. "The Influence of Invention and Discovery." In *Recent Social Trends,* by the President's Research Committee on Social Trends. New York: McGraw-Hill Book Co., 1934.

———. *On Culture and Social Change.* Chicago: University of Chicago Press, 1964.

Opinion Research Corporation. *The Intellectual's Challenge to the Corporate Executive.* Princeton, N.J.: Opinion Research Corporation, 1961.

———. *The Conflict Between the Management Mind and the Scientific Mind.* Princeton, N.J.: Opinion Research Corporation, 1959.

Organization for European Economic Cooperation. *The Organization of Applied Research in Europe*; proceedings of the conference held at Nancy, 11–13 October 1954. Vol. I.

Orth, Charles D., III. "The Optimum Climate for Industrial Research." In *Administering Research and Development: The Behavior of Scientists and Engineers in Organization,* edited by Charles D. Orth, III, Joseph C. Bailey, and Francis Wolek, Homewood, Ill.: Richard D. Irwin, Inc., 1964.

———, Joseph C. Bailey, and Francis Wolek, eds. *Administering Research and Development: The Behavior of Scientists and En-*

gineers in Organization. Homewood, Ill.: Richard D. Irwin, Inc., 1964.
Orth, Charles D., III, et al. *Managers and Scientists: Some Human Problems in Industrial Research Organizations.* Boston: Division of Research, Graduate School of Business Administration, Harvard University, 1963.
Peabody, R. L. *Organizational Authority: Superior Subordinate Relationships in Three Public Service Organizations.* New York: Atherton Press, 1964.
Pease, Hugh M. "Raising Productivity in Problem Solving Operations," *Personnel Administration,* vol. 27, no. 1 (January–February 1964), pp. 39–42.
Peck, Merton J. "Inventions in the Postwar American Aluminum Industry." In National Bureau for Economic Research, *The Rate and Direction of Inventive Activities.* Princeton, N.J.: Princeton University Press, 1962.
Pelz, D. C. "Motivation of the Engineering and Research Specialist," American Management Association, *General Management Series,* no. 186 (1957), pp. 25–46.
Perlman, Jacob. "Measurements of Scientific Research and Development." In *Operations Research in Research and Development,* edited by Burton V. Dean. New York: John Wiley & Sons, Inc., 1963.
Pierson, F. C. et al. *The Education of American Business.* New York: McGraw-Hill Book Co., 1959.
Pollack, F. *Automation: A Study of its Economic and Social Consequences.* New York: Frederick A. Praeger, Inc., 1957.
Porterfield, A. R. *Creative Factors in Scientific Research.* Durham, N.C.: Duke University Press, 1941.
Powell, Norman J. *Personnel Administration in Government.* Englewood Cliffs, N.J.: Prentice-Hall, Inc., 1956.
Presthus, Robert V. *The Organizational Society.* New York: Alfred A. Knopf, 1962.
Price, Don. *Government and Science.* New York: New York University Press, 1954.
Pye, Lucien, ed. *Communication and Political Development.* Princeton, N.J.: Princeton University Press, 1963.
Quinn, James B. "Top Management Guides for Research Planning." In *Technological Planning on the Corporate Level,* edited by

James R. Bright. Cambridge, Mass.: Harvard University Graduate School of Business, 1962.
———. *Yardstick of Industrial Research.* New York: Ronald Press Co., 1959.
Randle, C. Wilson. "Problems of R and D Management," *Harvard Business Review*, vol. 37, no. 1 (January–February 1959).
Reiss, A. J., Jr. *Occupations and Social Status.* New York: The Free Press, 1961.
Reissman, Leonard. "A Study of Role Conceptions in Bureaucracy," *Social Forces,* vol. 27, no. 3 (March 1949).
Research and Development Progress Reporting. Policy and Procedures Division, Air Force System Command, USAF, Washington, D.C., 1961.
Rice, A. K. *Productivity and Social Organizations: The Ahmedabad Experiment.* London: Tavistock Publications, 1958.
Riegel, John W. *Administration of Salaries and Intangible Rewards for Engineers and Scientists.* Ann Arbor: Bureau of Industrial Relations, University of Michigan, 1958.
———. *Collective Bargaining as Viewed by Unorganized Engineers and Scientists.* Ann Arbor: Bureau of Industrial Relations, University of Michigan, 1959.
Riggs, Fred W. "Agraria and Industria—Toward a Typology of Comparative Administration." In *Toward the Comparative Study of Public Administration,* edited by William J. Stiffin. Bloomington: Indiana University Press, 1959.
Roberts, E. B. *The Dynamics of Research and Development.* New York: Harper & Row, 1964.
Roethlisberger, F., and W. J. Dickson. *Management and the Worker.* Cambridge: Harvard University Press, 1939.
Rogers, Everett M. "Characteristics of Agricultural Innovators and Other Adopter Categories." In *Studies of Innovation and of Communication to the Public.* Institute of Communications Research, Stanford University, 1962.
Rubin, H. *Pensions and Employee Mobility in the Public Service.* New York: Twentieth Century Fund, 1965.
———. *Diffusion of Innovation.* New York: The Macmillan Co., 1962.
Russo, Francois. "Scientific and Technical Creation," *American Behavioral Scientist,* vol. 6, no. 5 (January 1961).
Salter, W. E. G. *Productivity and Technical Change.* New York: Cambridge University Press, 1960.

Sanders, Barkev S. "Some Difficulties in Measuring Inventive Activities." In National Bureau of Economic Research, *The Rate and Direction of Inventive Activities*. Princeton, N.J.: Princeton University Press, 1962.

Schiff, Ashley L. "Innovation and Administrative Decision Making: The Conservation of Land Resources," *Administrative Science Quarterly*, vol. 11, no. 1 (June 1966), pp. 1–30.

Schumpeter, Joseph A. *Capitalism, Socialism and Democracy*. New York: Harper & Brothers, 1950.

———. *The Theory of Economic Development*. Cambridge: Harvard University Press, 1951.

Scott, William G. *The Management of Conflict: Appeal Systems in Organizations*. Homewood, Ill.: Richard D. Irwin, Inc., 1965.

Seimer, S. J. *Suggested Plans in American Industry*. Syracuse: Syracuse University Press, 1959.

Shapiro, Martin. *Law and Politics in the Supreme Court*. New York: The Macmillan Co., 1964.

Shepard, Herbert. "Nine Dilemmas of Industrial Research," *Administrative Science Quarterly*, vol. 1 (December 1956), pp. 295–309.

———. "Superiors and Subordinates in Research," *Journal of Business*, vol. 29, no. 4 (October 1956), pp. 261–67.

———. "The Dual Hierarchy in Research." In *Administering Research and Development: The Behavior of Scientists and Engineers in Organization*, edited by Charles D. Orth, III, Joseph C. Bailey, and Francis Wolek. Homewood, Ill.: Richard D. Irwin, Inc., 1964.

Shils, Edward. "Demagogues and Cadres in the Political Development of the New States." In *Communication and Political Development*, edited by Lucien Pye. Princeton, N.J.: Princeton University Press, 1963.

Siffin, William J., ed. *Toward the Comparative Study of Public Administration*. Bloomington: Indiana University Press, 1959.

Silk, L. S. *The Research Revolution*. New York: McGraw-Hill Book Co., 1960.

Simon, Herbert A. "A Behavioral Model of Rational Choice," *Quarterly Journal of Economics*, vol. 69, no. 1 (February 1955), pp. 99–118.

———. *The New Science of Management Decision*. New York: Harper & Brothers, 1960.

Smith, D. B. *Technological Planning on the Corporate Level.* Cambridge: Division of Research, Graduate School of Business Administration, Harvard University, 1962.
Smith, Paul, ed. *Creativity: An Examination of Process.* New York: Hastings House, 1959.
Snow, C. P. *Science and Government.* New York: Mentor Books, 1962.
———. *The Two Cultures and a Second Look.* New York: Mentor Books, 1964.
Solow, Robert M. "Technological Change and the Aggregate Production Function," *Review of Economics and Statistics,* vol. 39 (August 1957), pp. 312–20.
Spengler, Joseph J. "Bureaucracy and Economic Development." In *Bureaucracy and Political Development,* edited by Joseph LaPalombara. Princeton, N.J.: Princeton University Press, 1963.
Stein, Morris I., and Shirley J. Heinze. *Creativity and the Individual.* Glencoe, Ill.: The Free Press, 1960.
———. *Survey of Psychological Literature in the Area of Creativity With View Toward Needed Research.* New York: Research Center for Human Relations and New York University Press, 1962.
Steiner, Gary A. "The Creative Organization." Selected Papers Number Ten. Chicago: Graduate School of Business, University of Chicago, 1962.
Stillman, Gabriel. *PERT: A New Management Planning and Control Technique.* New York: American Management Association, n.d.
Stires, David M., and Maurice M. Murphy. *PERT (Program Evaluation Review Technique) CPM (Critical Path Method).* Boston: Materials Management Institute, 1962.
Stone, Donald C. "Manning Tomorrow's Cities," *Public Administration Review,* vol. 23 (June 1963), pp. 99–104.
Storer, Norman W. "Some Sociological Aspects of Federal Science Policy," *American Behavioral Scientist,* vol. 6 (December 1962), pp. 27–30.
Stover, Carl F. *The Goverment and Science.* Santa Barbara: Center for the Study of Democratic Institutions, 1962.
Strausmann, W. P. *Risk and Technical Innovation.*
Strauss, A. L., and L. Rainwater. *The Professional Scientist: A Study of American Chemists.* Chicago: Aldine Publishing Co., 1962.

Studies of Innovation and of Communication to the Public. Studies in the Utilization of Behavioral Science. Vol. 2. Stanford: Institute for Communication Research, Stanford University, 1962.

Summerfield, J., and L. Thatcher, eds. *The Creative Mind and Method.* Austin: University of Texas Press, 1960.

Super, D. E., and P. B. Bachrack. *Scientific Careers and Vocational Development Theory.* New York: Bureau of Publications, Teachers College Press, Columbia University, 1957.

Taylor, C. W., ed. *Creativity: Progress and Potential.* New York: McGraw-Hill Book Co., 1964.

———, and F. Barron, eds. *Scientific Creativity: Its Recognition and Development.* New York: John Wiley & Sons, Inc., 1963.

Thayer, L. O. *Administrative Communication.* Homewood, Ill.: Richard D. Irwin, Inc., 1961.

Thompson, James D., and Arthur Tuden. "Strategies, Structures, and Processes of Organizational Decision." In *Comparative Studies in Administration.* Pittsburgh: University of Pittsburgh Press, 1959.

Thompson, S. *How Companies Plan.* New York: American Management Association, 1962.

Thompson, Victor A. "Bureaucracy and Innovation," *Administrative Science Quarterly,* vol. 10, no. 1 (June 1965), pp. 1–20.

———. "Bureaucracy in a Democratic Society." In *Public Administration and Democracy,* edited by Roscoe C. Martin. Syracuse, N.Y.: Syracuse University Press, 1965.

———. *Modern Organization.* New York: Alfred A. Knopf, Inc., 1961.

———. *The Regulatory Process in OPA Rationing.* New York: King's Crown Press, 1950.

Thurstone, L. *Creative Talent.* Chicago: University of Chicago Press, 1950.

Turner, A., and P. Lawrence. *Industrial Jobs and The Worker: An Investigation of Response to Task Attitude.* Cambridge: Graduate School of Business Administration, Harvard University Press, 1965.

U.S. Bureau of the Budget. *Measuring Productivity of Federal Government Organizations.* Washington: Government Printing Office, 1964.

Usher, A. P. *A History of Mechanical Inventions.* 1st ed. New York:

McGraw-Hill, Inc., 1929. Rev. ed. Cambridge: Harvard University Press, 1954.

Vaughan, Floyd L. *Economics of Our Patent System.* New York: The Macmillan Co., 1925.

Veblen, Thorstein. *The Engineers and the Price System.* New York: Viking Press, Inc., 1921.

Villers, Raymond. *Research and Development: Planning and Control.* New York: Financial Executives Research Foundation, 1964.

———. "The Scheduling of Engineering Research," *Journal of Industrial Engineering,* vol. 10, no. 6 (November–December 1959).

Walker, Nigel. *Morale in the Civil Service: A Study of the Desk Worker.* Edinburgh: The University Press, 1960.

Wallas, Graham. *The Art of Thought.* New York: Harcourt Brace, 1926.

Wallis, W. Allen. "Some Economic Considerations." In *Automation and Technological Change,* edited by John T. Dunlop. Englewood Cliffs, N.J.: Prentice-Hall, Inc., 1962.

Warner, W. Lloyd, *The Corporation in the Emergent American Soiety.* New York: Harper & Row, 1962.

Warner, W. Lloyd, and James Abegglen, *Big Business Leaders in America.* New York: Harper & Row, 1955.

———. *Industrial Man; Businessmen and Business Organizations.* New York: Harper & Row, 1959.

———. *Occupational Mobility in American Business and Industry.* Minneapolis: University of Minnesota Press, 1955.

Warner, W. Lloyd, Paul P. Van Riper, Norman H. Martin, and Orvis F. Collins. *The American Federal Executive.* New Haven: Yale University Press, 1963.

Weber, Max. *The Theory of Social and Economic Organization.* Trans. by A. M. Henderson and Talcott Parsons. Talcott Parsons, ed. New York: Oxford University Press, Inc., 1947.

Whisler, Thomas L. "Executives and Their Jobs—the Changing Organizational Structure." *Selected Papers Number Nine.* Chicago: Graduate School of Business, University of Chicago, 1962.

White, Fred A. *American Industrial Research Laboratories.* Washington: Public Affairs Press, 1961.

Whiting, C. S. *Creative Thinking.* Reinhold Publishing Corp., 1958.

Whyte, William H., Jr. *The Organization Man.* Garden City, N.Y.: Doubleday Anchor Books, Doubleday & Co., Inc., 1957.

Wiener, Norbert. *The Human Use of Human Beings: Cybernetics and Society*. Boston: Houghton Mifflin Co., 1950.
Wildavsky, Aaron. *The Politics of the Budgetary Process*. Boston: Little, Brown & Co., 1964.
Wilensky, Harold L. *Intellectuals in Labor Unions*. Glencoe, Ill.: The Free Press, 1956.
———. *Organizational Intelligence*. New York: Basic Books, Inc., 1967.
Wilkening, E. A. "The Communication of Ideas on Innovation in Agriculture." In *Studies of Innovation and of Communication to the Public*. Stanford: Institute of Communications Research, Stanford University, 1962.
Wilson, D. R. *The Direction of Research Establishments*. London: H.M. Stationery Office, 1957.
Wilson, Logan, ed. *Emerging Patterns in American Higher Education*. Washington: American Council on Education, 1965.
Wolfe, D. *Science and Public Policy*. Lincoln: University of Nebraska Press, 1959.
Wormald, F. L. *The Pugwash Experiment*. Washington: Association of American Colleges, 1958.
Zaleznik, A., and D. Moment. *Role Development and Interpersonal Competence: An Experimental Study of Role Performance in Problem Solving Groups*. Cambridge: Division of Research, Graduate School of Business Administration, Harvard University, 1963.
Zaleznik, A., C. R. Christenson, and F. J. Roethlisberger. *The Motivation, Productivity, and Satisfaction of Workers*. Cambridge: Division of Research, Graduate School of Business Administration, Harvard University, 1958.

Index

Achievement: level, 35, 45, 46, 97; norms, 13; *vs.* ascription, 91
Administrative power elite, 53, 55, 56, 83
Aggregative grouping, 25, 26, 72
Alcoa, 48
Aluminum industry, 48
American Society for Public Administration, 100
Anderson, Admiral David, 50
Antiballistic missile (ABM), 56, 57
Anxiety, 11, 19, 93, 98, 99
Appeals, 20, 51
Applied mathematicians, 53, 58, 99
Area-based organization, 90
Argyris, Chris, 68, 79
Army Reserves, 50
Aspiration level, 35, 44, 45, 46
Authority, 15, 16, 17, 19, 21, 23, 63, 64, 65, 89, 92, 93, 97, 99, 103, 104
Automation, 36, 39, 40, 44, 45, 91, 96, 99
Automotive industry, 39

Autonomy, 69, 84, 86, 100

Baltimore, 39
Bargaining, 16, 58, 101
Barnard, Chester, 101
Behavioral: revolution, 50; scientists, 49
Blame, 24, 26
Blue-collar worker, 18, 20, 96, 97
Brainstorming, 14
Bright, James, 36
Boguslaw, Robert, 53, 58
"Bootlegging," 85
Boundary, of an organization, 24, 82, 83, 84, 86, 87
Buck-passing, 24
Budgeting, 40, 41, 44, 48, 49, 83, 86
Buffalo, 39
Bureau of the Budget, 53, 58
Bureaucratic orientation, 21, 22

Capital: human, 94, 96; requirements, 4, 34, 39, 43, 98, 99
Case studies, 68
Centralization, 98, 99

Charisma, 28, 47
Chemical Abstracts, 66
Chemists, 66
Chicago, 39
Cincinnati, 39
Citation counts, 68
Claimants, 48, 59
Cleveland, 39
Coalition, 16, 30, 48
Codification of knowledge, 97
Comparative Administration, 62
Competition, 10, 11, 13, 18, 19, 45, 46, 47, 66, 86, 98
Compliance, 17
Compromise, 48, 57, 93, 101
Computers, 55, 98, 100
Commitment of resources, 32, 35
Communication, 9, 13, 17, 22, 24, 31, 47, 67, 72, 76–78, 79, 85, 89, 90, 94, 101
Conflict, 16, 17, 51
Conformity, 14, 18, 19, 69, 86, 93, 99
Congress, 48, 51, 53
Consensus, 57
Control, 14, 15, 17, 20, 29, 30, 31, 32, 36, 37, 43, 44, 45, 46, 58, 59, 83, 84, 85, 86, 93, 98, 99, 100, 103; centralized, 32, 43, 75; monocratic, 46; pluralistic, 46
Control-oriented management, 31, 58
Controller, 83, 102, 103
Cosmopolitanism, 73
Cost-benefit analysis, 58, 100, 104
CPM (Critical Path Method), 37
Creativity, 9, 10, 11, 12, 14, 35, 46, 63, 86, 106
Cyert Richard, 30

Data processing, 4, 32, 55, 99, 101, 102, 103, 104
Decentralization, 43, 98
Decision makers, 4, 98
Decision making, 3, 4, 7, 22, 27–28, 36, 43, 46, 49, 50, 55, 61, 89, 100
Declaration of Independence, 38
Department of Defense, 56, 58, 60
Department of Justice, 56
Departmentalization, 25, 80, 101
Desk class, 20, 21, 25, 26, 32, 72, 92, 93, 94; management as, 22
Detroit, 39
Development, 14, 34, 35, 38, 46, 65, 66, 84, 96, 98, 99
Diffusion studies, 68
Discipline, 16, 17
"District Officer," 85
Diversity of inputs, 10, 12, 25, 26, 31, 32, 72, 73, 80
Docility, 17, 18
"Dropouts," 97
Dual salary ladders, 42
Duplication, 16, 17, 24, 25, 33, 47, 67, 73

Econologicians, 53, 54, 55, 58, 61, 83, 99, 100
Economists, 49
Economy and efficiency, 48, 53, 58
Education, 6, 33, 68, 69, 74, 90; continuing, 95, 97
Efficiency, 4, 5, 30, 32, 41, 46, 48, 53, 58, 66, 100, 103
Einstein, Albert, 40
Employee satisfaction, 6, 64, 82

Engineers, 19, 28, 30, 32, 36, 39, 40, 42, 47, 65, 66, 69, 70, 94
Engineering, 28, 65
Engineering Index, 66
Entrepreneur, 4, 5
Establishment, 86
Esteem, 11, 13, 19, 21
Estimates, 37, 85
Etzioni, Amatai, 62
Evaluation, 11, 27, 29, 30, 36, 40, 47, 69, 79, 92, 93, 104
Expert, 12, 13, 68, 79

Failure, 4, 5, 10, 26, 44
Federal Aviation Agency, 51
Federal government, 38, 49, 53, 57, 58
Feedback, 16
Financial Executives Research Foundation, 44
Ford, Henry, 29
Ford Motor Company, 39
Freedom, 10, 11, 12, 19, 47, 77, 78, 99, 100, 101, 102

Gantt, Henry, 52
Generalist, 12, 13, 81, 85
Goals: group, 13; organizational, 22, 26, 29, 30, 33, 48, 53, 54, 65, 75, 84; personal, 13, 27
Great Britain, 85
Group problem-solving, 19
Groups, 9, 11, 12, 13, 14, 25, 27, 29, 65, 73, 80, 81, 83, 89, 99

Handbook of Organizations, 61, 62
Hand-over transactions, 86
Harvard Weapons Acquisition Project, 46

Hawthorne investigations, 53, 58
Hersberg, Frederick, 105
Hierarchy, 13, 15, 17, 18, 19, 21, 27, 33, 43, 47, 50, 98, 102
Historians, 49
House Armed Services Committee, 50

Identification, 24, 70, 71; organizational, 86; professional, 86
Ideology, 3, 7, 29, 68, 84
Illinois Institute of Technology, 56
Income, 11; tax, 59, 101
Insecurity, 19
Integrative grouping, 25, 26, 72
Interface relations, 82–88
Internal Revenue Service, 50
Invention, 34, 35, 40, 73, 79, 90, 104
Inventor, 11

Jewkes, John, et al., 46
Job satisfaction, 78
Johnson, President Lyndon B., 53, 58
Joint Chiefs of Staff, 81
Joy in work, 17

Kennedy, President John F., 53
Klein, Burton, 47
Kornhauser, William, 71
Kuhn, Thomas, 83

Labor, 20, 34, 36, 92, 94, 96
Latent functions, 15, 59
Lawyers, 69
Leadership, 27
Legitimacy, 16, 19, 23, 35, 43
Leisure, 91, 106

LeMay, General Curtis, 50
Light bulb industry, 45
Lindblom, Charles, 33

McNamara, Defense Secretary Robert, 53, 56, 58
Management, 4, 6, 18, 20, 22, 29, 31, 33, 36, 37, 42, 43, 44, 45, 48, 49, 54, 55, 58, 59, 64, 83, 84, 92, 93, 96, 100, 102, 103, 104, 105, 106; profession, 22
"Managerialism," 102, 103, 104
Manifest functions, 59–60
March, James, 30
Marginal analysis, 33, 34, 48
Maximization, 29, 30, 43, 45
Means-end map, 43
Measurements, of organization, 63–79
Meier, Richard, 91
Mill, John Stuart, 47, 100, 101
Milton, John, 47, 100, 101
Minimax risk, 43
Mobility, 21, 85, 90, 92
Mohr, Lawrence, 68
Money, 17, 18, 69, 86
Monocratic organization, 15, 16, 19, 20, 21, 23, 25, 26, 27, 33, 46, 51, 86, 87, 92, 95, 102
Monocratic stereotype, 15, 16, 17, 28, 47, 50, 51
Monopoly, 46, 48
Morale, 21, 28, 44
Morse, Representative Bradford, 56
Motivation, 10, 42, 54, 68, 95, 97, 103

National Guard, 50

National Science Foundation, 38, 39, 49, 66, 67, 94
Negotiation, 48, 58, 93, 101
Nelson, Senator Gaylord, 56
Nelson, Richard, 37
Neo-Taylorites, 36, 53, 54, 55, 56, 58, 59, 60
"New science of management decision," 99, 100
New utopians, 53, 54, 55, 59
Norms, 13, 14, 61, 86
Nürnberg trials, 51

Obsolescence, 74, 95, 96
Oligopoly, 45, 46
Open society, 18
Operations research, 53
Optimism, 40, 41, 42, 52, 95, 106
Organization men, 21, 71
Organization theory, 3, 4, 7, 61, 62, 63, 71, 80, 102
Orwell, George, 102
Overlapping, 16, 17, 24, 25, 33, 47
Overrequirement, 32
Overspecification of human resources, 20, 21, 25, 31, 32
Owner, of the organization, 16, 17, 29, 30, 31, 32, 33, 35, 36, 37

Page, General Jerry, 50
Parochialism, 24, 25, 73, 87, 94, 95
Patents, 66, 67, 79
Payoffs, 44, 66
Peers, 11
Pentagon, 50, 56
Personality, 7, 63, 85, 99
Personnel administration, 42, 50, 93, 94, 105, 106

INDEX

PERT (Program Evaluation Review Technique), 37, 55
Ph.D.s, 42, 70
Philadelphia, 39
Physical Review, The, 67
Physical sciences, 52, 56
Physicists, 66
Pittsburgh, 39
Plato, 106
Platonism, 55
Pluralism, 16, 23, 47, 48, 86, 93
Political system, in organization, 19, 22, 23, 25, 87
Position classification, 93–94
Power, 11, 17, 22, 27, 69, 85, 93, 99
PPB (Planning–Programming–Budgeting), 53, 54, 59, 60
Pratt and Whitney, 48
Predictability, 5, 6, 15, 17, 30, 31
Pre-entry preparation, 18, 21, 92, 105
President of the United States, 48, 81
Privacy of the individual, 104
Probability, 26, 34, 40, 43; subjective, 40, 43
Problem solving, 11, 12, 14, 33, 44, 45, 48, 52, 76, 80, 81
Procedures writers, 58
Production, 7, 10, 30, 34, 36, 39, 40, 42, 46, 47, 63, 68, 75, 78, 82, 99, 103; function, 34; ideology, 19, 29, 30, 35; orientation, 32, 33, 84, 85; values, 6, 20, 31, 32, 38, 54, 64, 74, 83, 103
Professional association, 87, 105; growth, 19, 28, 31, 69, 93; identification, 70, 71; peers, 19, 21, 69, 93, 95; renegade, 22; values, 31, 69, 71, 85, 86
Professionalism, 68, 69, 72, 74, 92, 93, 94, 95, 104, 105
Professionals, 69, 70, 71, 72, 80–81, 84, 85, 92, 95, 104, 105
Profit, 6, 35, 44
Programmed activities, 31, 33, 47, 48, 75, 91
Project: management, 47; organization, 65
Promotion, 21, 42, 93
Psychologists, 49, 61
Public administration, 48, 49, 52, 60, 102
Public Administration Review, 49
Publications, 66, 67, 79
Purposiveness, 74–75

Rand Corporation, 58
R and D (Research and Development), 22, 23, 24, 34, 35, 37, 38, 39, 40, 41, 44, 45, 46, 47, 48, 49, 60, 66, 67, 82, 84, 87, 91
Rationality: classical model, 7, 33, 36, 40, 54; economic, 7, 33, 35, 36, 37, 38, 41, 42, 46, 58; individual, 26, 37; organizational, 26, 43; synoptic, 33, 54, 58, 59
Research, 6, 8, 34, 37, 38, 39, 40, 41, 44, 49, 53, 61, 62, 63, 64, 65, 69, 79–83, 84, 85, 87; applied, 34, 42, 69; basic, 34, 35, 36, 38, 41, 42, 45, 46, 66, 69; industrial, 36, 39
Researchers, 36, 37, 66, 84, 85, 86
Responsibility, 12, 16, 24, 25, 26, 27, 43, 94, 97, 104
Restructuring, 27

Revenue Act of 1954, 66
Rewards, 11, 13, 17, 18, 19, 21, 22, 23, 24, 25, 26, 42, 71, 92, 93
Riggs, Fred, 62
Roles, 3, 4, 12, 13, 15, 17, 33, 61, 63, 72, 103, 105
Rolls-Royce, 48
Rules of thumb, 55

Sabbaticals, 97
St. Louis, 39
Satisficing, 45
Scherer, Frederic, 46
Schumpeter, Joseph, 45, 46
Scientific management, 48, 52, 53
"Scientism," 56, 57, 99
Scientists, 28, 39, 41, 42, 49, 52, 65, 69, 70, 94, 99
Seaborg, Glen, 56
Search, in decision-making, 7, 10, 22, 31, 33, 34, 35, 72
Secretary of Defense, 81
Security, 10, 11, 12, 19, 37, 40, 41, 43, 99, 100
Selective forgetting, 86
Selective inattention, 86
Self-actualization, 105, 106
Self-fulfilling prophecy, 40
Senate Constitutional Rights Subcommittee, 50, 51
Serendipity, 10
Shoe industry, 45
Shortages of skilled personnel, 42
Simon, Herbert, 99
Slack, 30, 35, 42, 43, 44, 45, 46, 87
Smith, Adam, 30
Smith's pins effect, 30, 31, 32
Snow, C. P., 52
Social class, 18, 92

Social problems, 52, 56, 57, 58, 99
Social science, 52, 53, 58, 61
Social scientists, 49, 50, 53, 54
Social system, 3, 7, 9, 14, 15, 59, 83
Specialist, 12, 18, 80, 81
Specialization, 94, 96, 98
Specialty, 13, 97
Stability, 15, 85, 96
Standard Metropolitan Statistical Areas, 39
Status, 17, 18, 21, 22, 27, 30, 64, 68, 69, 86, 89, 93
Status quo, 22, 23, 84
Stratification, 13
Stress, organizational, 45
Success, 13, 18, 20, 25, 26, 40, 41, 42, 44, 52, 97, 102
Suggestion box systems, 6, 25, 26, 32, 73
Survival, 6, 82, 100
Sutton, Major General W. J., 50
"System approach," 56
System designers, 56

Taylor, Frederick, 36, 48, 53, 54, 55, 56, 58, 60
Technical development, 34
Technological change, 5, 41, 68, 98, 99, 103
"Technological fix," 56
Technology, 4, 15, 31, 39, 47, 92, 96, 104
Thompson, James, 99
Tizzard Committee, 52
Tool, organization as, 16, 17, 19, 29, 31, 100
Trade-offs, 6, 7, 64

Traditionalism, 91
Training, 74

Udy, Stanley, 62
Uncertainty, 6, 10, 16, 17, 34, 38, 46, 47, 60, 67
Unions, 20
United States Civil Service Commission, 94
Universalism vs. particularism, 91
University of Michigan, 68

Vandenberg Air Force Base, 51
Veblen, Thorstein, 12, 81

Veto, 19, 20
Villers, Raymond, 44

Wallas, Graham, 9
Washington, D.C., 39, 59
Waste, 33
Weapon, organization as, 16, 29, 100
Weapons development, 14, 37, 46, 60
Weber, Max, 15, 28, 33, 34, 47, 50, 51, 95, 102
White-collar worker, 20, 42, 96
Work, upgrading of, 92, 95, 96
World War II, 38